Mathematics and its Applications

Series Editor: G. M. BELL, Professor of Mathematics, King's College (KQC), University of London

Mathematics and its applications are now awe-inspiring in their scope, variety and depth. Not only is there rapid growth in pure mathematics and its applications to the traditional fields of the physical sciences, engineering and statistics, but new fields of application are emerging in biology, ecology and social organisation. The user of mathematics must assimilate subtle new techniques and also learn to handle the great power of the computer efficiently and economically.

The need of clear, concise and authoritative texts is thus greater than ever and our series will endeavour to supply this need. It aims to be comprehensive and yet flexible. Works surveying recent research will introduce new areas and up-to-date mathematical methods. Undergraduate texts on established topics will stimulate student interest by including applications relevant at the present day. The series will also include selected volumes of lecture notes which will enable certain important topics to be presented earlier than would otherwise be possible.

In all these ways it is hoped to render a valuable service to those who learn, teach, develop and use mathematics. *For full series list see end of book.*

STABILITY OF PARALLEL GAS FLOWS

STABILITY OF PARALLEL GAS FLOWS

BHIMSEN K. SHIVAMOGGI
Associate Professor of Mathematics
University of Central Florida, USA *and*
Institute of Mathematical Sciences, Madras, India

ELLIS HORWOOD LIMITED
Publishers · Chichester

Halsted Press: a division of
JOHN WILEY & SONS
New York · Chichester · Brisbane · Toronto

First published in 1986 by
ELLIS HORWOOD LIMITED
Market Cross House, Cooper Street, Chichester, West Sussex, PO19 1EB, England

The publisher's colophon is reproduced from James Gillison's drawing of the ancient Market Cross, Chichester.

Distributors:
Australia and New Zealand:
Jacaranda-Wiley Ltd., Jacaranda Press,
JOHN WILEY & SONS INC.
GPO Box 859, Brisbane, Queensland 4001, Australia
Canada:
JOHN WILEY & SONS CANADA LIMITED
22 Worcester Road, Rexdale, Ontario, Canada
Europe and Africa:
JOHN WILEY & SONS LIMITED
Baffins Lane, Chichester, West Sussex, England
North and South America and the rest of the world:
Halsted Press: a division of
JOHN WILEY & SONS
605 Third Avenue, New York, NY 10158, USA

British Library Cataloguing in Publication Data
Shivamoggi, Bhimsen K.
Stabilibity of parallel gas flow. —
(Ellis Horwood series in maths and its applications)
1. Gas flow 2. Compressibility
I. Title
533'.28 QC163

ISBN 0-7458-0001-7 (Ellis Horwood Limited)
ISBN 0-470-20287-4 (Halsted Press)

Printed in Great Britain by R. J. Acford, Chichester

Contents

'The final aim of natural science is to resolve itself into mathematics.'

H. von Helmholtz

In memory of my late revered teacher, Gurunath Hulyalkar

Preface

In recognition of the importance of phenomena associated with gas flow at high speeds, this book is concerned with the study of the stability of parallel flows in a gas with respect to infinitesimal disturbances. A predominantly analytical approach is adopted, and no detailed comparison has been made between the theory and the experimental results. No detailed account is given (and one is not possible within the framework of the linear theory) of the mechanism of transition of laminar flow to turbulence, even though the latter remains the chief stimulus to research in hydrodynamic stability.

The present book is not intended to be a survey of the literature, but an introductory account of compressibility effects in stability of parallel flows. As such, a complete listing of references is not presented. References are usually given only when needed to note the fundamental advances.

Orlando, 1985 B. K. Shivamoggi

ACKNOWLEDGEMENTS

I would like to express my gratitude to Professor W.V.R. Malkus who introduced me to this subject, and to several authors, particularly, Professors P.G. Drazin, L.N. Howard and C.C. Lin, whose writings have contributed a great deal to my education on this subject. I wish to express my thanks to Mr Ellis Horword, the

Publisher, for his immense help and cooperation in this whole project. My thanks are due to my wife Jayashree for her gracious support of this undertaking, and to Sue Turpie, Sue Watson and Joan Dunham for their help in the preparation of the manuscript.

B.K. Shivamoggi

1

Theory of Hydrodynamic Stability

The objectives of the theory of hydrodynamic stability are:

(i) to determine whether a given flow (or a given class of flows) is ultimately unstable for sufficiently large Reynolds numbers (a simple, general criterion, which would readily classify the flow profiles according to their stability is desirable);

(ii) to determine the minimum critical Reynolds number at which instability begins (it may be easier to find sufficient conditions for stability than to find the condition for passage from stability to instability);

(iii) to understand the physical mechanism underlying the phenomena by giving theoretical interpretations and experimental confirmations of the results obtained from mathematical analysis.

The most extensive discussion of hydrodynamic stability has been the study of inertial instabilities in parallel flows. The physical cause of instability of parallel flows is much subtler in contrast to several other kinds of instability such as thermal instability, free-surace – and interface – instability, gravitational instability.

It has now been generally recognised that turbulence is the more natural state of viscous fluid flows, and laminar motion is very likely to occur when the Reynolds number is so low because the disturbances are likely to be damped out by viscosity. The stability of laminar flow relative to infinitsimally-small disturbances is the problem of hydrodynamic stability. This in turn is governed by the properties of the fluid, the nature of the mean flow, and the nature of the disturbance.

In the case of a dynamical system with a finite number of degrees of freedom, a definition of stability of equilibrium or of steady motion can be easily given, because the motion is determined by the initial conditions. But the case of a continuous medium renders this difficult in that the determination of the motion requires a knowledge not only of the condition at time $t = 0$, but also the boundary conditions for $t > 0$.

Mathematically the problem consists of following the subsequent career of a hydrodynamically-possible disturbance which is superposed on the basic steady flow. If the disturbances vanish as $t \to \infty$, the motion is stable; otherwise it is unstable.

In order to study the behaviour of disturbances, one must follow the solution of a system of nonlinear partial-differential equations, which, in general, is very difficult. Instead one assumes that the disturbances are small so that the equations governing the disturbance may be linearised, i.e. the terms quadratic or higher in the disturbances and their derivatives may be neglected. The resultant linear and homogeneous system of equations contains time t only through derivatives so that the solutions containing an exponential time factor e^{ct} may be expected. The boundary conditions for the disturbances (which require the vanishing of quantities such as the disturbance velocities at the boundaries) are normally homogeneous; consequently, we have a characteristic-value problem with c as the parameter. If all the characteristic values of c have negative real parts, the motion is stable with respect to infinitesimal disturbances. If some of the characteristic values of c have positive real parts, the motion is said to be unstable.

However, it must be noted that such a linear instability does not necessary lead to turbulence; it could lead to another more stable state of laminar motion. Linear theory is sufficient if one is interested only in the onset of instability.

Because of the increasing importance of phenomena associated with gas flow at high speeds, Lees and Lin (1946) extended the investigation to cover the case of a gas, taking into account the effects of compressibility and heat conductivity. The experimental results of Laufer and Vrebalovich (1960), which indicated the existence of Tollmien waves in a laminar compressible boundary layer, have given credence to the theory of stability of the laminar compressible boundary layer as a useful guide to some of the important factors that control transition to turbulent flow. Fig. 1 shows the close agreement of the neutral-stability curve found experimentally by Laufer and Vrebalovich (1960) with the numerically-calculated one by Mach (1965). The physical causes of turbulence in the boundary layer and the mixing layer are yet to be understood, but they appear to be associated with the presence of finite disturbances which may be introduced from the free stream (example: ambient turbulence), or be caused by surface imperfections in the case of a boundary layer, and interfacial shear deformation in the case of a mixing layer. In the absence of such sources of

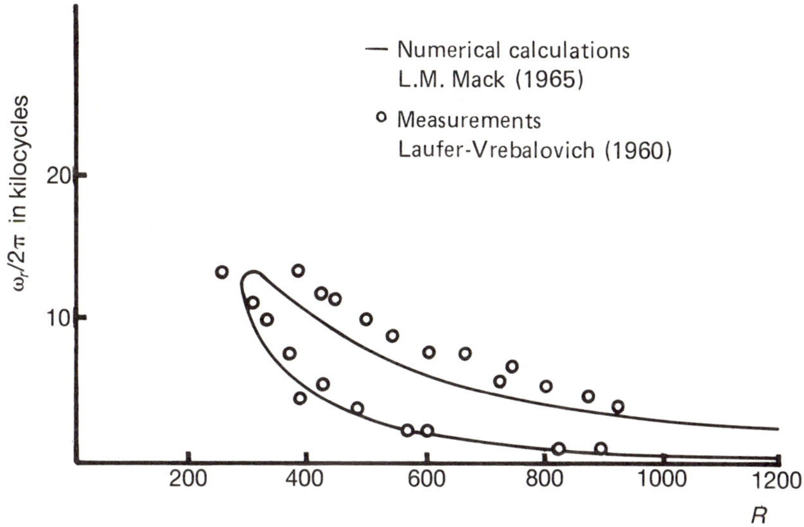

Fig. 1. Neutral-stability curve of a compressible boundary layer at Mack 2.2 (insulated wall). (From Mack (1965) as given in R. Betchov and W.O. Criminale, *Stability of Parallel Flows*, Academic Press, 1967.)

disturbances it is believed that transition to turbulence may be provoked by sufficiently-amplified disturbances initially of the type shown by the linear theory to cause instability of laminar compressible parallel flow. It may be noted that the free shear flows, such as the flow in a mixing layer become unstable at very low Reynolds numbers, the case of which has been presumed to be an inviscid-instability mechanism that does not operate in the boundary-layer flows (see Fig. 2).

The equations of flow are linearised with respect to small disturbances of the basic parallel flow. We adopt here the method of normal modes, in which each disturbance is resolved into dynamically-independent wave components. A linear characteristic-value problem is then posed to determine the typical component. The characteristic-value problem is singular, and the singularity admits solutions with discontinuous derivatives and continuous spectrum of characteristic values in addition to well-behaved solutions with a discrete spectrum. Two-dimensional disturbances are considered for reasons of mathematical convenience. However, the disturbances are known to develop three-dimensionally in an essential manner in the process of transition.

Though a more permanent outline of the subject has emerged as a result of eminent studies by Helmholtz (1868), Kelvin (1887), Rayleigh (1892), Orr (1907), Sommerfeld (1908), Heisenberg (1924), Tollmien (1935), Synge (1933), Lin (1945), and many others, the subject of hydrodynamic stability of parallel flows remains well known for its controversial nature. The effect of viscosity in parallel flows is one such example. Intuition would suggest that viscosity is essentially dissipative in nature, and therefore provides a mechanism for damping out the disturbances. If one further observes that instability generally occurs at large Reynolds numbers, one might reason that the principal features of the

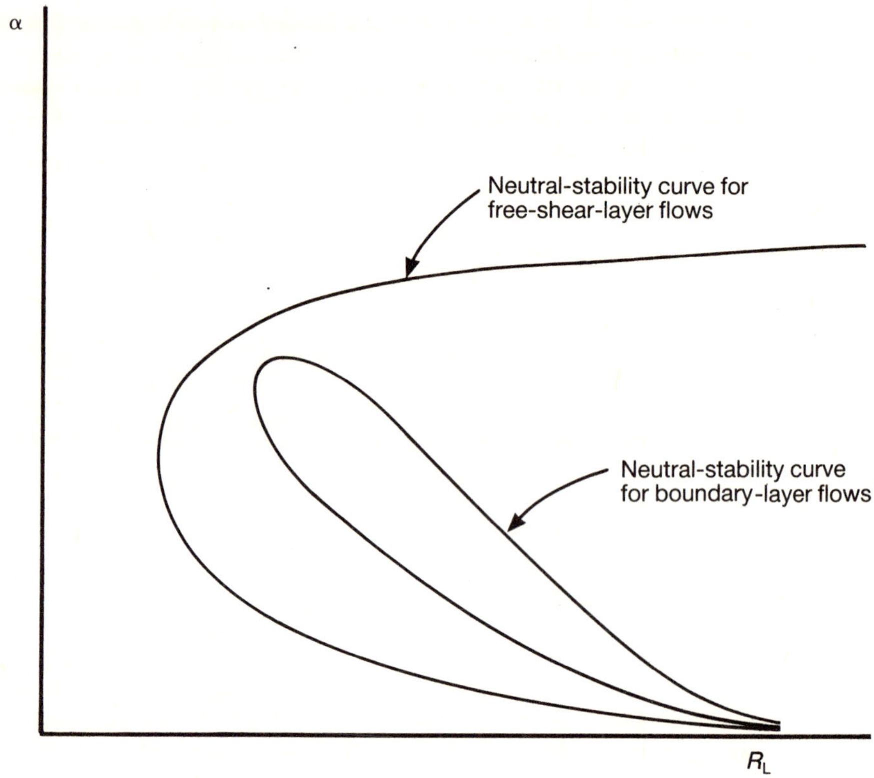

Fig. 2

mechanism of instability may be obtained by first neglecting the viscous effects, and then incorporating them later as a stabilising influence.

But experiments show that viscous effects can also serve as a cause of instability. In the case of plane Poiseuille flow, Rayleigh's criterion would certainly indicate stability in the absence of viscosity. The theory does show that the flow is stable with respect to disturbances of a given wavelength, provided the Reynolds number is high enough. However, for wavelengths beyond a certain lower limit, there is always a finite range of Reynolds numbers for which the motion is unstable. This is so since by some means, viscosity also promotes the transfer of energy from the basic laminar flow to the disturbance, which is thereby amplified. Physically, the dual action of viscosity is connected with the fact that its effects are diffusive as well as dissipative in character, and consequently it can assist in spreading any local concentration of vorticity to neighbouring parts of the fluid. This diffusive action materialises since the theory shows that the supposed regions of concentrated vorticity do exist in the disturbed flow. (Indeed, Rayleigh believed that an investigation in which viscosity is altogether ignored may not be applicable to the limiting case of a viscous fluid when the viscosity approaches zero.)

Let us go through a brief review of the stability theory for incompressible flows prior to embarking on a detailed discussion on the stability of compressible flows.

2

Stability of Parallel Incompressible Flows

(1) INTRODUCTION

The inviscid theory of stability of steady two-dimensional flows with parallel streamlines was started by Helmholtz (1868), Kelvin (1887) and Rayleigh (1880). Rayleigh (1880) deduced his celebrated theorem on the role of inflexion points in the velocity profile, which was even verified by Reynolds (1883) independently in his celebrated experiments on transition to turbulence. The viscous theory began with the work of Orr (1907) and Sommerfeld (1908), who derived the famous equation now bearing their names. Various asymptotic methods of analysis of this equation were advanced by Heisenberg (1924), Tollmien (1935), and Lin (1945). The basic premise of the Tollmien–Lin theory, that the stability or instability of infinitesimal disturbances depends only on their frequency or wavelength and on the Reynolds number, was confirmed by the experiments of Schubauer and Skramstad (1947), showing that the transition to turbulence was preceded by this instability. A detailed up-to-date account of this problem is available in the monograph by Drazin & Reid (1981).

(2) THE ORR–SOMMERFELD EQUATION

Consider two-dimensional flows of an incompressible fluid between two parallel planes. Let x denote the distance parallel to the flow, y the distance perpendicular to the planes, and z the distance normal to the flow, but coplanar with the bounding planes. Let (u, v, w) be the corresponding velocity components. Let the various flow variables be nondimensionalised using a reference length L and a reference velocity U and mass density ρ. The Navier–Stokes equations are then

$$\frac{\partial u}{\partial t}+u\frac{\partial u}{\partial x}+v\frac{\partial u}{\partial y}+w\frac{\partial u}{\partial z}=-\frac{\partial p}{\partial x}+\frac{1}{R_{\mathrm{L}}}\nabla^2 u \tag{1}$$

$$\frac{\partial v}{\partial t}+u\frac{\partial v}{\partial x}+v\frac{\partial v}{\partial y}+w\frac{\partial v}{\partial z}=-\frac{\partial p}{\partial y}+\frac{1}{R_{\mathrm{L}}}\nabla^2 v \tag{2}$$

$$\frac{\partial w}{\partial t}+u\frac{\partial w}{\partial z}+v\frac{\partial w}{\partial y}+w\frac{\partial w}{\partial z}=-\frac{\partial p}{\partial z}+\frac{1}{R_{\mathrm{L}}}\nabla^2 w \tag{3}$$

$$\frac{\partial u}{\partial x}+\frac{\partial v}{\partial y}+\frac{\partial w}{\partial z}=0 \tag{4}$$

where

$$\nabla^2\equiv\frac{\partial^2}{\partial x^2}+\frac{\partial^2}{\partial y^2}+\frac{\partial^2}{\partial z^2},\quad R_{\mathrm{L}}\equiv\frac{UL\rho}{\mu}$$

μ being the viscosity of the fluid.

Consider perturbations about a mean parallel flow $\bar{u}(y)$ parallel to the x-axis:

$$u=\bar{u}+u',\quad v=v',\quad w=w',\quad p=\bar{p}+p'. \tag{5}$$

Using (5), and linearising the perturbation, equations (1)–(4) give

$$\frac{\partial u'}{\partial t}+\bar{u}\frac{\partial u'}{\partial x}+v'\frac{\mathrm{d}\bar{u}}{\mathrm{d}y}=-\frac{\partial p'}{\partial x}+\frac{1}{R_{\mathrm{L}}}\nabla^2 u' \tag{6}$$

$$\frac{\partial v'}{\partial t}+\bar{u}\frac{\partial v'}{\partial x}=-\frac{\partial p'}{\partial y}+\frac{1}{R_{\mathrm{L}}}\nabla^2 v' \tag{7}$$

$$\frac{\partial w'}{\partial t}+\bar{u}\frac{\partial w'}{\partial z}=-\frac{\partial p'}{\partial z}+\frac{1}{R_{\mathrm{L}}}\nabla^2 w' \tag{8}$$

$$\frac{\partial u'}{\partial x}+\frac{\partial v'}{\partial y}+\frac{\partial w'}{\partial z}=0 \tag{9}$$

In the theory of normal modes, one looks for solutions of the form

$$\{u',v',w',p'\}=\{u'(y),v'(y),w'(y),p'(y)\}\,\mathrm{e}^{\mathrm{i}(\alpha x+\beta z-\alpha ct)} \tag{10}$$

where α and β are real, and $c=c_{\mathrm{r}}+\mathrm{i}c_{\mathrm{i}}$, so that the disturbances grow or decay in time like $\mathrm{e}^{\alpha c_{\mathrm{i}}t}$. Using (10), and eliminating p', equations (6)–(9) give

$$(\bar{u}-c)(\mathrm{D}^2-\gamma^2)v'-v'\,\mathrm{D}^2\bar{u}=-\frac{\mathrm{i}}{\alpha R_{\mathrm{L}}}(\mathrm{D}^2-\gamma^2)^2 v' \tag{11}$$

$$(\bar{u}-c)(\beta u'-\alpha w')-\frac{\mathrm{i}\beta}{\alpha}v'\mathrm{D}\bar{u}=-\frac{\mathrm{i}}{\alpha R_{\mathrm{L}}}(\mathrm{D}^2-\gamma^2)(\beta u'-\alpha w') \qquad (12)$$

$$\mathrm{i}\alpha u'+\mathrm{i}\beta w'+\mathrm{D}v'=0 \qquad (13)$$

where

$$\mathrm{D}\equiv\frac{\mathrm{d}}{\mathrm{d}y}, \quad \gamma^2\equiv\alpha^2+\beta^2$$

The boundary conditions for a flow bounded by rigid planes at $y = 0, 1$ are

$$y=0,1:\ u', v', w'=0 \qquad (14a)$$

or using (13), one has

$$y=0,1:\ v', \mathrm{D}v'=0 \qquad (14b)$$

From equation (12), note that it is possible to make the following transformation:

$$\beta u'-\alpha w'=\gamma\tilde{u}', \quad \alpha R_{\mathrm{L}}=\gamma\tilde{R}_{\mathrm{L}} \qquad (15)$$

to arrive at an equivalent two-dimensional problem ($\beta = 0$). Thus, the three-dimensional evolution of a disturbance at a Reynolds number R_{L} is equivalent to the two-dimensional evolution of the disturbance at a lower Reynolds number $\tilde{R}_{\mathrm{L}}$. In the case of neutral disturbances ($c_{\mathrm{i}} = 0$), therefore, the minimum critical Reynolds number occurs for a two-dimensional disturbance, so that in seeking a sufficient criterion for instability one may consider only the two-dimensional disturbances, which greatly simplifies the analysis. This is called Squire's Theorem (Squire, 1933).

Using Squire's Theorem, and introducing

$$u'=\frac{\partial\phi}{\partial y}, \quad v'=-\mathrm{i}\alpha\phi \qquad (16)$$

where ϕ is the amplitude of a stream function

$$\psi=\phi(y)\,\mathrm{e}^{\mathrm{i}\alpha(x-ct)} \qquad (17)$$

equation (11) gives the Orr–Sommerfeld equation (Orr, 1907; Sommerfeld, 1908)

$$(u-c)(\phi''-\alpha^2\phi)-u''\phi=-\frac{\mathrm{i}}{\alpha R_{\mathrm{L}}}(\phi''''-2\alpha^2\phi''+\alpha^4\phi) \qquad (18)$$

where the bar over u has been dropped without causing further ambiguity, and the primes denote differentiation with respect to y.

The boundary conditions (14) becomes

$$y=0,1:\quad \phi, \phi'=0 \qquad (19)$$

The characteristic-value relation for this problem then takes the form

$$F(\alpha, c, R_L) = 0 \tag{20}$$

In the usual temporal-stability problem, in which one considers the growth of a disturbance in time, one takes α and R_L to be real and given. The characteristic-value relation above then defines a discrete set of characteristic values c_j ($j = 1, 2, \ldots$). If one writes

$$c = c_r(\alpha, R_L) + ic_i(\alpha, R_L)$$

then the curves of neutral stability are given by $c_i(\alpha, R_L) = 0$.

(3) INVISCID PROBLEM

In the inviscid limit ($R_L \to \infty$), (18) and (19) become

$$(u - c)(D^2 - \alpha^2)\phi - D^2 u \,.\, \phi = 0 \tag{21}$$

$$y = 0, 1: \quad \phi = 0 \tag{22}$$

Note that since in the inviscid limit, the order of the equation for ϕ has dropped by two, one needs to drop the boundary condition of no-slip at the walls. Further, if ϕ is an eigenfunction with eigenvalue c for some α, then ϕ^* (ϕ^* being the complex conjugate of ϕ) is an eigenfunction with eigenvalue c^* for the same α. Thus, to each growing mode there exists a corresponding damped mode. It is convenient to adopt the convention of recognising only the growing mode and ignoring the damped mode.

Next, note that unless

$$(u'')_{u=c} = 0$$

the point $y = y_c$, where $u = c$, in the complex y-plane is a regular singularity of equation (21). Otherwise, the inviscid system (21) and (22) would have a well-defined characteristic-value problem. The singular nature of equation (21) is also one of the chief difficulties in constructing uniformly-valid asymptotic solutions to equation (18). Besides, if $c_i \leqslant 0$, it turns out that a regular solution of equation (21) satisfying the boundary conditions (22) does not exist if $0 < y_c < 1$.

In the inviscid problem, u is an arbitrary function of y.

(i) Inviscid solutions

Corresponding to $\alpha = 0$, equation (21) has two solutions:

$$\phi_{10} = u - c$$

$$\phi_{20} = (u - c) \int_0^y \frac{dy}{(u - c)^2} \tag{23}$$

Seeking solutions to equation (21), then, in ascending powers of α^2 (Heisenberg, 1924):

$$\phi_1 = \phi_{10} + \alpha^2\phi_{11} + \ldots$$

$$\phi_2 = \phi_{20} + \alpha^2\phi_{21} + \ldots \tag{24}$$

and noting that equation (21) can be written as

$$\left[\left(\frac{\phi_{n1}}{u-c}\right)'(u-c)^2\right]' = (u-c)\,\phi_{no}\,; \quad n = 1, 2, \ldots$$

one obtains

$$\phi_{n1} = (u-c)\int_0^y \frac{dy}{(u-c)^2}\int_0^y (u-c)\,\phi_{no}\,dy; \quad n = 1, 2, \ldots \tag{25}$$

Consideration of the initial-value problem (Case, 1960) or the viscous problem (Lin, 1945) shows that the path of integration from $y = 0$ to y is continued around the critical point $y = y_c$ (when c is real) from below the real axis in the complex y-plane when $u_c' > 0$, and vice versa.

Let us next obtain two linearly-independent solutions in the neighbourhood of the point $y = y_c$. First write $(u - c)$ as a Taylor-series expansion around the point $y = y_c$,

$$u - c = u_c' \cdot \eta + \frac{u_c''}{2!}\eta^2 + \ldots$$

where

$$\eta = y - y_c.$$

Then, equation (21) becomes

$$\eta\phi'' - \left(\frac{u_c''}{u_c'} + \alpha^2\eta\right)\phi = 0. \tag{26}$$

Since $\eta = 0$ is a regular singular point of equation (26), seek a solution of equation (26) of the form (Frobenius' Method)

$$\phi_1 = \eta^m \sum_{n=0}^{\infty} a_n\eta^n, \quad \eta > 0 \tag{27}$$

Using (27), equation (26) gives

$$\sum_{n=-1}^{\infty} a_{n+1}(m+n)(m+n+1)\,\eta^{m+n} - \frac{u_c''}{u_c'}\sum_{n=0}^{\infty} a_n\eta^{m+n}$$

$$-\alpha^2\sum_{n=1}^{\infty} a_{n-1}\eta^{m+n} = 0$$

from which

$$[a_0(m-1)m]\,\eta^{m-1} + \left[a_1(m+1)m - \frac{u_c''}{u_c'}a_0\right]\eta^m +$$

$$+\sum_{n=1}^{\infty}\left[a_{n+1}(m+n)(m+n+1) - \frac{u_c''}{u_c'}a_n - \alpha^2 a_{n-1}\right]\eta^{m+n} = 0 \tag{28}$$

Equating the coefficients of each power of η to zero, one obtains, first, the indicial equation

$$a_0(m-1)m = 0 \tag{29}$$

from which, since $a_0 \neq 0$,

$$m = 1 \quad \text{or} \quad 0$$

Next, one obtains from (28)

$$a_1 = \frac{a_0}{2}\frac{u_c''}{u_c'}$$

Thus,

$$\phi_1(\eta) = a_0\left[\eta + \frac{1}{2}\frac{u_c''}{u_c'}\eta^2 + \ldots\right] \tag{30}$$

Now, note that since the two roots of the indicial equation (29) differ by an integer, the other linearly-independent solution is given by

$$\phi_2(\eta) = a\phi_1(\eta)\ln\eta + \eta^0\left[\sum_{n=0}^{\infty} b_n\eta^m\right] \tag{31}$$

Using (31), equation (26) gives

$$\eta\left[a\left(a_0\frac{u_c''}{u_c'} + \ldots\right)\ln\eta + 2aa_0\left(1 + \frac{u_c''}{u_c'}\eta + \ldots\right)\frac{1}{\eta} +\right.$$

$$\left.-aa_0\left(\eta + \frac{1}{2}\frac{u_c''}{u_c'}\eta^2 + \ldots\right)\frac{1}{\eta^2} + \sum_{n=0}^{\infty} b_n\, n(n-1)\,\eta^{n-2}\right] +$$

$$-aa_0\frac{u_c''}{u_c'}\left(\eta + \frac{1}{2}\frac{u_c''}{u_c'}\eta^2 + \ldots\right) - \frac{u_c''}{u_c'}\sum_{n=0}^{\infty} b_n\eta^n +$$

$$-\alpha^2\sum_{n=0}^{\infty} b_n\eta^{n+1} = 0$$

Equating the coefficient of the term η^0 to zero, one obtains

$$a a_0 - \frac{u_c''}{u_c'} b_0 = 0$$

Thus,

$$\phi_2(\eta) = \left(\frac{u_c''}{u_c'} \ln \eta \right) \phi_1(\eta) + \left[1 + \sum_{n=1}^{\infty} b_n \eta^n \right] \tag{32}$$

Note that because of the singular nature of $\phi_2(\eta)$, at least in the neighbourhood of y_c (where it is discontinuous and multi-valued), it cannot be a uniformly-valid asymptotic approximation to any solution of equation (18). Further, one has to decide the proper branch to be taken in connection with the logarithmic singularity at y_c. If one assumes the neutral modes as the limit of unstable modes (Lin, 1955), then one has the following analytical continuation for $\ln \eta$:

$$\ln \eta \sim \begin{cases} \ln \eta, & \eta > 0 \\ \ln |\eta| - \mathrm{i}\pi, & \eta < 0 \end{cases}$$

(ii) **Discontinuities in the mean flow**

For an inviscid fluid, $u(y)$ can be an arbitrary function of y, and even discontinuities are admissible. Thus, if u or u' is discontinuous at y_0, say, then the pressure must be continuous at the material interface with mean position at $y = y_0$. Therefore, noting that

$$(u - c)\phi'' - u''\phi = \{(u - c)\phi' - u'\phi\}' = (p)'$$

one has

$$[(u - c)\phi' - u'\phi] = 0 \tag{33}$$

at $y = y_0$, where the square brackets denote the jump of their contents at such a discontinuity. Also, the normal velocity of the fluid must be continuous at the material interface. Let this interface be represented by

$$y = y_0 + \xi(x, t) \tag{34}$$

where

$$\xi(x, t) \sim \mathrm{e}^{\mathrm{i}\alpha(x - ct)}$$

Then, one has

$$v' = \frac{\partial \xi}{\partial t} + u \frac{\partial \xi}{\partial x} = \mathrm{i}\alpha(u - c)\xi \tag{35}$$

Noting that $v' = \alpha\phi$, one obtains

$$\left[\frac{\phi}{u-c}\right] = 0 \tag{36}$$

at $y = y_0$.

Explicit solutions of the characteristic-value problem of (21) and (22) are difficult to find in practice, for smoothly-varying functions $u(y)$. However, when $u(y)$ is a piecewise-linear function, the situation is tractable, and one can then join up such functions by using conditions (33) and (36). As an illustration, consider two uniform fluids in relative horizontal motion of velocities U_1, U_2 separated by a horizontal boundary at $y = 0$. One has

$$y \gtrless 0 : \phi''_{1,2} - \alpha^2 \phi_{1,2} = 0 \tag{37}$$

from which, noting that

$$y \to \pm\infty : \phi_{1,2} \to 0$$

and using (36), one obtains

$$y > 0 : \phi_1 = A(U_1 - c)\,\mathrm{e}^{-\alpha y}$$

$$y < 0 : \phi_2 = A(U_2 - c)\,\mathrm{e}^{\alpha y} \tag{38}$$

Using (38), (33) gives the dispersion relation

$$(U_1 - c)^2 + (U_2 - c)^2 = 0 \tag{39}$$

from which

$$n = -\alpha c = \frac{1}{2}(U_1 + U_2) \pm \frac{\mathrm{i}\alpha}{2}\,|U_1 - U_2| \tag{40}$$

so that such an interface is always unstable.

It is possible to effect some inhibition of the instability of such an interface by introducing a vorticity appropriately in the two streams. Let us introduce a constant vorticity ω (along the z-direction) in the two streams. The velocity distribution in the two streams in the steady state is then given by

$$y \gtrless 0 : u_{1,2} = U_{1,2}\left(1 - \frac{\omega y}{U_{1,2}}\right) \tag{41}$$

Using (38) and (41), (33) gives the dispersion relation

$$(U_1 - c)^2 + (U_2 - c)^2 - \alpha\omega(U_1 - U_2) = 0 \tag{42}$$

from which

$$n = -\alpha c = \frac{1}{2}(U_1 + U_2) \pm \frac{1}{2}\left[\frac{2\omega}{\alpha}(U_1 - U_2) - (U_1 - U_2)^2\right]^{1/2} \tag{43}$$

which shows that, if the ambient shear is opposite in sign to that of the interfacial shear, it is possible to inhibit the instability of such an interface. In order to understand this result physically, let us recall that the instability of an interface between two streams is caused by the relative motion of the fluids tangential to the interface: in other words, an interfacial shear. One may then expect that any agent such as an ambient shear, that opposes the interfacial shear, effectively reduces the relative motion of the fluids at the interface and hence would inhibit the instability of the latter.

(iii) **Local and global necessary conditions for the existence of unstable modes**

THEOREM: *If the flow possesses a nonneutral mode of disturbance with a finite wavelength, the mean-velocity profile has an inflexion point at some point* $y = y_s$, *where* $0 \leqslant y \leqslant 1$, *i.e. it satisfies a local necessary condition (Rayleigh, 1880)*

$$(u'')_{y=y_s} = 0$$

Furthermore, in the case of a neutral mode, one must have $u(y_s) = c$, $c, = c_r$.

Proof: Multiply equation (21) by ϕ^*, the complex conjugate of ϕ, and integrate from 0 to 1; then, using (22), one obtains

$$-\int_0^1 [|\phi'|^2 + \alpha^2 |\phi|^2]\, dy = \int_0^1 \frac{u''}{u-c} |\phi|^2 dy \tag{44}$$

The imaginary part of (44) gives

$$c_i \int_0^1 \frac{u''}{|u-c|^2} |\phi|^2 dy = 0 \tag{45}$$

If $c_i \neq 0$, it follows that u'' must change sign at one or more points in the flow field. Assuming that u'' is continuous for $0 \leqslant y \leqslant 1$, there must be at least one point in the interval (0, 1) where one has

$$(u'')_{y=y_s} = 0 \tag{46}$$

Next, in order to show that in the case of a neutral mode one must have $u(y_s) = c\ (c = c_r)$, multiply equation (21) by ϕ^*, and subtract from the resulting equation its complex conjugate, and then there follows

$$\frac{d}{dy}(\phi'\phi^* - \phi^{*\prime}\phi) - (c - c^*) \frac{u''}{|u-c|^2} |\phi|^2 = 0 \tag{47}$$

Introducing the 'Wronskian'

$$W \equiv \frac{i}{2}(\phi\phi^{*\prime} - \phi'\phi^*) \tag{48}$$

equation (47) gives

$$\frac{dW}{dy} - c_i \left(\frac{u''}{|u-c|^2} \right) |\phi|^2 = 0 \tag{49}$$

If one lets $c_i \to 0$ in equation (49), one obtains results corresponding to neutral disturbances. Thus,

$$W = \text{constant} \tag{50}$$

except possibly at the critical point y_c (where $u = c$, $c = c_r$) which may now lie in the real interval (0, 1). If it does, W has a discontinuity at the point y_c. Integrate equation (49) from y_c^- to y_c^+, then one obtains

$$W(y_c^+) - W(y_c^-) = \int_{u_c^-}^{u_c^+} \frac{u'' \, |\phi|^2 c_i \, du}{u' \, |u-c|^2} \tag{51}$$

Thus, as $c_i \to 0$, (51) gives

$$[W]_c \equiv W(y_c^+) - W(y_c^-) = \pi \frac{u_c''}{u_c'} |\phi_c|^2 \tag{52}$$

provided in this limit $c_i > 0$ and $u_c^+ > u_c^-$. (Here the subscript c refers to the values at the point y_c.) Equation (52) shows that, if the mean-flow profile is monotonic, a neutral mode exists when all the jumps $[W]_c = 0$; i.e. $u'' = 0$ at all points where $u = c$ ($c = c_r$). In this case, equation (21) has regular solutions which are physically possible in the absence of viscous effects.

Let us next deduce a global necessary condition (Fjortoft, 1950) for the existence of a nonneutral mode. The real part of (44) gives

$$\int_0^1 (u - c_r) \frac{u''}{|u-c|^2} |\phi|^2 \, dy = - \int_0^1 [|\phi'|^2 + \alpha^2 |\phi|^2] \, dy \tag{53}$$

Adding

$$(c_r - u_s) \int_0^1 \frac{u''}{|u-c|^2} |\phi|^2 \, dy = 0$$

to the left-hand side of equation (53), one obtains

$$\int_0^1 \frac{u'' \cdot (u - u_s)}{|u-c|^2} |\phi|^2 \, dy = - \int_0^1 [|\phi'|^2 + \alpha^2 |\phi|^2] \, dy \tag{54}$$

From equation (54), one obtains a global necessary condition for the existence of a nonneutral mode

$$u'' \cdot (u - u_s) < 0 \tag{55}$$

where, for $0 \leqslant y \leqslant 1$,

$$(u'')_{y=y_s} = 0$$

and

$$u_s = u(y_s). \tag{56}$$

In particular if $u(y)$ is a monotonic function of y, and u'' vanishes only once in the flow field, $0 \leqslant y \leqslant 1$, a combined local and global necessary condition for instability is that

$$u'' \cdot (u - u_s) \leqslant 0 \tag{57}$$

throughout the flow field $0 \leqslant y \leqslant 1$, with equality valid only where $y = y_s$. Physically, this means that the magnitude of vorticity $|u'|$ at the point of inflexion y_s must be a maximum.

Let us now give a counter-example to the sufficiency of the above necessary condition for instability. Consider a mean flow given by (Drazin and Howard, 1966)

$$y_1 \leqslant y \leqslant y_2 : u = \sin y$$

for which

$$D^2 u = 0 \quad \text{when } y = y_s = n\pi; \qquad n = 0, \pm 1, \pm 2, \ldots$$

If there is no value y_s in the interval (y_1, y_2), the flow is certainly stable by the above necessary condition for instability. If there is at least one value, one may suppose $y_s = 0$ without loss of generality, so $y_1 \leqslant 0 \leqslant y_2$. Since $u(y)$ is odd, for each characteristic function $\phi(y)$ with a characteristic value c, there is a characteristic function $\phi(-y)$ with a characteristic value $\tilde{c} = -c_r + ic_i$, for the same α. The uniqueness of the characteristic solution then implies that $c_r = 0$. Thus, one has for the odd solution from equation (21):

$$D^2\phi + (1 - \alpha^2)\phi = 0$$

$$y = y_1, y_2 : \phi = 0$$

from which

$$\phi = \sin \frac{n\pi(y - y_1)}{y_2 - y_1}$$

where

$$\alpha = \left[1 - \frac{n^2 \pi^2}{(y_2 - y_1)^2}\right]^{1/2}$$

for each positive integer $n < (y_2 - y_1)/\pi$. It then follows that the flow is unstable if $(y_2 - y_1) > \pi$, but stable otherwise, although the point of inflexion lies at $y = 0$ in the field of flow. This problem also illustrates the stabilising effect of the walls on the flow.

(iv) Howard's Semi-circle Theorem

Let us next deduce a semi-circle theorem for nonneutral disturbances. First, construct a function $g(y)$ which is supposed to be regualr in the interval (0, 1):

$$g(y) = \frac{\phi(y)}{u(y) - c} \tag{58}$$

then, equation (21) becomes

$$[(u-c)^2 g']' - \alpha^2 (u-c)^2 g = 0 \tag{59}$$

Multiply equation (59) by g^*, g^* being the complex conjugate of g, and integrate between (0, 1); then one obtains

$$\int_0^1 (u-c)^2 \left\{|g'|^2 + \alpha^2 |g|^2\right\} \mathrm{d}y = 0 \tag{60}$$

THEOREM: *If $c_i \neq 0$, c_r must lie within the range of u (Synge, 1933).*

Proof: The real and imaginary parts of equation (60) give

$$\int_0^1 \left\{(u - c_r)^2 - c_i^2\right\} \left\{|g'|^2 + \alpha^2 |g|^2\right\} \mathrm{d}y = 0 \tag{61}$$

and

$$c_i \int_0^1 (u - c_r) \left\{|g'|^2 + \alpha^2 |g|^2\right\} \mathrm{d}y = 0 \tag{62}$$

Equation (62) shows that if $c_i \neq 0$, c_r must lie within the range of u.

THEOREM: *The complex wave velocity for a self-excited mode lies inside the semi-circle in the upper half of the c-plane which has the range of u for its diameter along the real axis (Howard, 1961).*

Proof: For $c_i \neq 0$, one has from (62),

$$\int_0^1 u\left\{|g'|^2 + \alpha^2 |g|^2\right\} \mathrm{d}y$$

$$= c_r \int_0^1 \left\{|g'|^2 + \alpha^2 |g|^2\right\} \mathrm{d}y \tag{63}$$

Using (63), (61) becomes

$$\int_0^1 u^2\{|g'|^2+\alpha^2|g|^2\}\,\mathrm{d}y$$

$$=\int_0^1 (c_r^2+c_i^2)\{|g'|^2+\alpha^2|g|^2\}\,\mathrm{d}y \tag{64}$$

Now, using (63) and (64) in the inequality

$$0\geqslant\int_0^1 (u-u_{\min})(u-u_{\max})\{|g'|^2+\alpha^2|g|^2\}\,\mathrm{d}y$$

one obtains

$$\int_0^1 \{|g'|^2+\alpha^2|g|^2\}\{(c_r^2+c_i^2)-(u_{\min}+u_{\max})c_r$$

$$+u_{\min}u_{\max}\}\,\mathrm{d}y\leqslant 0 \tag{65}$$

from which there follows

$$c_r^2+c_i^2-(u_{\min}+u_{\max})c_r+u_{\min}u_{\max}\leqslant 0$$

or

$$[c_r-\tfrac{1}{2}(u_{\min}+u_{\max})]^2+c_i^2\leqslant[\tfrac{1}{2}(u_{\max}-u_{\min})]^2 \tag{66}$$

which shows that the complex wave velocity for an unstable mode must lie inside the semi-circle in the upper half of the complex c-plane which has the range of u for its diameter along the real axis.

(v) Sufficient conditions for the existence of unstable and neutral modes

Once the global necessary condition in the form

$$0<y<1:-\frac{u''}{u(y)-u(y_s)}\geqslant 0$$

holds, then one turns to the question of sufficient conditions for the existence of unstable and neutral modes. Toward this, one first exhibits the variational character of the inviscid problem for monotonic symmetrical type of mean flow, and then uses the Sturm–Liouville Theorem to show that there exists a neutrally-stable characteristic solution with $c=c_r=u_s$, and then constructs unstable solutions that are contiguous with the neutral modes, i.e. $c\to u_s$ as $c_i\to 0^+$.

THEOREM: *Regular neutral modes given by*

$$\phi=\phi_s,\quad \alpha=\alpha_s,\quad c=c_s=u_s \tag{67}$$

are possible if the mean-flow profiles are monotonic, and if one has in the flow field

$$(u'')_{y=y_s} = 0$$

(Friedrichs, 1942).

Proof: First write equation (21) as

$$\phi'' + K(y)\phi + \lambda\phi = 0 \tag{68}$$

where

$$\lambda = -\alpha^2, \quad K(y) = -\frac{u''}{u - c_s}$$

Let there by a point $y = y_s$ such that

$$(u'')_{y=y_s} = 0 \quad \text{and} \quad (u')_{y=y_s} \neq 0$$

In addition, let $u(y) > 0$ for $0 < y < 1$ and assume that $K(y)$ is a regular function for $0 < y < 1$. Then, equation (68) is a real nonsingular equation which makes up a Sturm–Liouville problem with the boundary conditions (22). This characteristic-value problem possesses an infinite sequence of characteristic values. In order to show that there is a characteristic value λ for which a neutral mode exists with $c = u(y_s)$, where $c = c_r$, one has to identify λ with $-\alpha^2$, i.e. one must show that at least one of the characteristic values of equation (68) is negative.

Multiply equation (68) by ϕ^*, and integrate from 0 to 1, and using the boundary conditions (22), one obtains

$$\lambda = \frac{\int_0^1 \{|\phi'|^2 - K|\phi|^2\} \,\mathrm{d}y}{\int_0^1 |\phi|^2 \,\mathrm{d}y} \tag{69}$$

This is a characteristic-value problem for the parameter $\lambda = -\alpha^2$ and is associated with the variational principle

$$\delta \int_0^1 \{|\phi'|^2 - K|\phi|^2\} \,\mathrm{d}y = 0$$

subject to

$$\int_0^1 |\phi|^2 \,\mathrm{d}y = \text{constant} \tag{70}$$

Thus, if ϕ really satisfies equation (68) as well as the boundary conditions (22), then λ is determined from (69) by noting that λ is stationary if ϕ is given a slight variation $\delta\phi$ satisfying only the boundary conditions (22). If $\delta\lambda$ represents the change in λ corresponding to $\delta\phi$, then one has

$$\delta\lambda = \frac{2I \int_0^1 (\phi'\delta\phi' - K\phi\delta\phi)\,dy - 2J \int_0^1 \phi\delta\phi\,dy}{I^2} \tag{71}$$

where

$$I \equiv \int_0^1 |\phi|^2 dy, \quad J \equiv \int_0^1 \{|\phi'|^2 - K|\phi|^2\}\,dy$$

One obtains from (71),

$$-\frac{I}{2}\delta\lambda \doteq \int_0^1 \{\phi'' + K\phi + \lambda\phi\}\,dy \tag{72}$$

Since ϕ satisfied equation (68), one obtains

$$\delta\lambda = 0 \tag{73}$$

Conversely, if $\delta\lambda$ vanishes for all variations $\delta\phi$ satisfying the boundary conditions and possessing a derivative which, like ϕ, is square-integrable, then (72) shows that the corresponding function ϕ satisfies equation (67). (73) merely shows that the characteristic value λ is stationary; one may readily show that λ is a minimum.

Next we deduce the existence of an unstable mode in the neighbourhood of the neutral mode, i.e. an unstable mode whose limit as $c_i \to 0^+$ is the neutral mode. That the neutral solution has a neighbouring unstable solution will be shown below by considering the dependence of c on $\lambda = -\alpha^2$ in the characteristic-value problem of equation (21) (Lin, 1945). Recalling that there is a characteristic function ϕ_s corresponding to the characteristic value $c = c_s$, $-\alpha_s^2 = \lambda_s$, one has

$$\phi_s'' - \frac{u''}{u - c_s}\phi_s + \lambda_s\phi_s = 0 \tag{74}$$

Multiply equation (21) by ϕ_s, equation (64) by ϕ, and subtract, so that

$$(\phi_s\phi' - \phi_s'\phi)' + (\lambda - \lambda_s)\phi\phi_s - \left(\frac{u''}{u - c} - \frac{u''}{u - c_s}\right)\phi\phi_s = 0 \tag{75}$$

Integrating between (0, 1), one obtains

$$(\lambda - \lambda_s)\int_0^1 \phi\phi_s\,dy = \int_0^1 \left(\frac{u''}{u - c} - \frac{u''}{u - c_s}\right)\phi\phi_s\,dy$$

which in the limit $\lambda \to \lambda_s$, $c \to c_s$, and $\phi \to \phi_s$ gives

$$\left(\frac{d\lambda}{dc}\right)_{\alpha=\alpha_s}\int_0^1 |\phi_s|^2 dy = \underset{c\to c_s}{\mathrm{Lt}} \int_0^1 \frac{u''}{u - c_s}\,\frac{|\phi_s|^2}{u - c}\,dy \tag{76}$$

The right-hand side of equation (76) is

$$\int_0^1 \frac{|\phi_s|^2}{u-c}\left(-\frac{u''}{u-c_s}\right) dy = \int_0^1 \frac{\left(-\dfrac{u''}{u-c_s}\right)(u-c_r)}{|u-c|^2} |\phi_s|^2 \, dy$$

$$+ \mathrm{i}c_i \int_0^1 \frac{\left(-\dfrac{u''}{u-c_s}\right)}{|u-c|^2} |\phi_s|^2 \, dy$$

The limit of the real part becomes the principal value of the integral

$$\int_0^1 \frac{\left(\dfrac{u''}{u-c_s}\right)}{u-c_s} |\phi_s|^2 \, dy$$

The imaginary part tends to the limit

$$\pi \left[-\frac{u''}{u_s'(u-c_s)}\right] |\phi_s|^2 > 0$$

as $c_i \to 0^+$. Equation (76) then gives an equation of the form

$$-\frac{\mathrm{d}\lambda}{\mathrm{d}c} = A + \mathrm{i}B, \quad B > 0$$

or

$$-\frac{\mathrm{d}c}{\mathrm{d}\lambda} = \frac{\mathrm{d}c}{\mathrm{d}\alpha^2} = \frac{A - \mathrm{i}B}{A^2 + B^2} \tag{77}$$

If α^2 decreases slightly, (77) shows that c_i becomes positive, which proves the existence of an unstable disturbance in the neighbourhood of the neutral disturbance.

(vi) The initial-value-problem approach

Since in the problem of hydrodynamic stability, one really wants to know whether an initial perturbation grows in time, strictly speaking, one should pose an initial-value problem. A notable merit of the initial-value-problem approach is that one cannot miss the modes associated with a continuous spectrum of characteristic values, unlike that in a normal-mode analysis (Case, 1960; Dikii, 1961). In order to illustrate this, consider a plane Couette flow between parallel planes at $y = 0$ and 1, with

$$u(y) = y \tag{78}$$

(a) The normal-mode approach

One has in the normal-mode approach, for this case,

$$\phi'' - \alpha^2 \phi = 0 \tag{79}$$

$$y = 0, 1 : \phi = 0 \tag{80}$$

the solution of which is

$$\phi \equiv 0 \tag{81}$$

This does not mean, however, that an inviscid plane Couette flow has no non-trivial solutions. Because, in going from equation (21) to equation (79), one has eliminated the modes corresponding to a continuous spectrum of characteristic values. These can be recovered by posing an initial-value problem.

(b) The initial-value-problem approach

Fourier transforming in x, and Laplace transforming in t as

$$\bar{\phi}(\alpha, y, s) = \int_0^{\infty} e^{-st} dt \int_{-\infty}^{\infty} e^{-i\alpha x} \psi(x, y, t) \, dx \tag{82}$$

one obtains from the original equations

$$\bar{\phi}'' - \alpha^2 \bar{\phi} = \frac{\phi_0'' - \alpha^2 \phi_0}{s + i\alpha y} \tag{83}$$

$$y = 0, 1 : \bar{\phi} = 0 \tag{84}$$

where

$$t = 0 : \phi = \phi_0 \tag{85}$$

Upon inverting the Laplace transform, equation (83) gives

$$\phi'' - \alpha^2 \phi = e^{-i\alpha y t} (\phi_0'' - \alpha^2 \phi_0) \tag{86}$$

Since there are no nonzero ϕ_0, such that

$$\left.\begin{aligned} \phi_0'' - \alpha^2 \phi_0 &= 0 \\ y = 0, 1 : \phi_0 &= 0 \end{aligned}\right\} \tag{87}$$

the right-hand side in equation (86) cannot be identically zero. What is giving one relief now is exactly what causes one dismay in the normal-mode approach! The solutions of equation (86) correspond to continuous characteristic values c, running between 0 and 1.

(4) VISCOUS PROBLEM

Because the situations of interest correspond to large values of the Reynolds number, the asymptotic methods of analysis should play an important role in the theory. We shall now construct asymptotic solutions to equation (18) using a WKBJ method (Lin, 1945; Reid, 1965). Seek solutions to equation (18) of the form

$$\phi = e^{\int g(y)\,dy} \tag{88}$$

so that equation (18) gives

$$(u-c)\left\{(g^2+g')-\alpha^2\right\}-u'' = -\frac{i}{\alpha R_L}\left\{g^4+6g^2g'+3g'^2+4gg''+g'''+ - 2\alpha^2(g^2+g')+\alpha^4\right\} \tag{89}$$

One may solve equation (89) by expanding $g(y)$ in powers of (αR_L) as

$$g(y) = (\alpha R_L)^{1/2} g_0(y) + g_1(y) + (\alpha R_L)^{-1/2} g_2(y) + \ldots \tag{90}$$

so that equation (89) gives

$$(u-c)\,g_0^2 = -i g_0^4 \tag{91}$$

$$(u-c)\,(g_0' + 2g_0 g_1) = -i(4g_0^3 g_1 + 6g_0^2 g_0') \text{ etc.} \tag{92}$$

Therefore,

$$g_0 = \pm\sqrt{i(u-c)} \tag{93}$$

$$g_1 = -\frac{5}{2}\frac{g_0'}{g_0} \quad \text{etc.} \tag{94}$$

Thus,

$$\phi(y) \approx C_{1,2}(u-c)^{5/4} \exp \pm \int_0^y \sqrt{i\alpha R_L(u-c)}\,dy \tag{95}$$

Note that the WKBJ solutions (95) are not valid in a domain of the complex y-plane that contains the critical point $u = c$. The determination of the correct branch so as to achieve the analytical continuation of (95) across the point $u = c$ would be such that (95) yields a valid asymptotic solution of equation (18) all along the path. If one takes a path from 0 to 1 in the complex y-plane below the critical point y_c, one is then always in a region in the complex y-plane where the above asymptotic solutions are valid (Lin, 1945).

In order to determine the location of the point y_c as a function of the sign of c_i in the complex y-plane, let us write $y = y_r + iy_i$, and analytically continue the velocity field $u(y)$ in the complex y-plane. Then $u = u(y)$ will be an analytical function. One has $u(y) \to u(y_r)$ as $y_i \to 0$ and the equation $u(y) = c$ must have a solution, namely $u(y_c) = c$. In order to determine y_c, let $y_c = y_{r_c} + iy_i$. Separating then the relation

$$u(y_{r_c} + iy_i) = c$$

into real and imaginary parts, one has

$$u(y_{r_c} + iy_i) = u_r + iu_i = c_r + ic_i$$

from which

$$u_r = c_r, \quad u_i = c_i$$

From small y_i, one may write the latter as

$$c_i = u_i(y_{r_c}) + \frac{\partial u_i}{\partial y_i}(y_{r_c}) \cdot y_i$$

Using the Cauchy–Riemann conditions

$$\frac{\partial u_r}{\partial y_r} = \frac{\partial u_i}{\partial y_i}$$

one then has

$$c_i = u_i(y_{r_c}) + \frac{\partial u_r}{\partial y_r}(y_{r_c}) \cdot y_i$$

Since $u_i \to 0$ as $y_i \to 0$, one has the result that if $u_c' > 0$ and $c_i > 0$, the point y_c lies above the real axis in the complex y-plane. Likewise, if $u_c' > 0$ and $c_i < 0$, then the point y_c lies below the real axis in the complex y-plane.

Let us now deduce two viscous solutions which are valid in a complex neighbourhood of the critical point y_c. First, note that the quantity $1/R_L$ multiplies the highest derivative in equation (18), so that a straightforward expansion in terms of the parameter $1/R_L$ will not be uniformly valid. An alternative method is to choose a small parameter ε related to $(\alpha R_L)^{-1/3}$ and make a transformation

$$y - y_c = \varepsilon\eta, \quad \varepsilon = (\alpha R_L u_c')^{-1/3} \tag{96}$$

Expanding $(u - c)$ in powers of $(y - y_c)$ and putting

$$\phi = \chi_0 + \varepsilon\chi_1 + \varepsilon^2\chi_2 + \ldots \tag{97}$$

one obtains from equation (18):

$$i\frac{d^4\chi_0}{d\eta^4} + \eta\frac{d^2\chi_0}{d\eta^2} = 0 \tag{98}$$

$$i\frac{d^2\chi_1}{d\eta^4} + \eta\frac{d^2\chi_1}{d\eta^2} = \frac{u_c''}{u_c'}\chi_0 - \frac{1}{2}\frac{u_c''}{u_c'}\eta^2\frac{d^2\chi_0}{d\eta^2} \tag{99}$$

etc.

The four basic solutions of equation (98) are

$$\chi_0^{(1)} = 1$$

$$\chi_0^{(2)} = \eta$$

$$\chi_0^{(3)} = \int_{\infty}^{\eta} \mathrm{d}\eta \int_{\infty}^{\eta} \eta^{1/2} H_{1/3}^{(1)} \left[\tfrac{2}{3} (\mathrm{i}\eta)^{3/2}\right] \mathrm{d}\eta$$

$$\chi_0^{(4)} = \int_{-\infty}^{\eta} \mathrm{d}\eta \int_{-\infty}^{\eta} \eta^{1/2} H_{1/3}^{(2)} \left[\tfrac{2}{3} (\mathrm{i}\eta)^{3/2}\right] \mathrm{d}\eta \qquad (100)$$

where $H_{1/3}^{(1)}$ and $H_{1/3}^{(2)}$ denote Hankel functions of order 1/3. The solutions $\chi_0^{(1)}$ and $\chi_0^{(2)}$ correspond to the terms of lowest powers of η in the inviscid solutions ϕ_1 and ϕ_2 of (28) and (32). On the other hand, $\chi_0^{(3)}$ and $\chi_0^{(4)}$ are viscous solutions. It can be shown by using the asymptotic properties of the Hankel functions for large values of η that $\chi_0^{(3)}$ tends to zero for large positive η, and $\chi_0^{(4)}$ tends to zero for large negative η. If one assumes that u_c' is positive, ε is positive, so that η is negative at the wall (assuming the wall to be at $y = 0$) and positive on the other side of y_c. Consequently, $\chi_0^{(3)}$ is the physically-realistic solution in representing the secondary boundary layer at the wall.

Thanks to the idea of a composite approximation which is commonly used in connexion with the method of matched asymptotic approximations, one may continue the foregoing approximations to form composite approximations which are valid in domains that do not shrink to zero as $\varepsilon \to 0$. The WKBJ solutions are valid provided $|y - y_c| \gg \varepsilon$. Thus, if $|\varepsilon| \ll |y - y_c|$ and $-\frac{7}{6}\pi < \mathrm{Ph}\,(y - y_c) < \frac{5}{6}\pi$, then one has from (95):

$$\phi \sim C_2 \left[u_c' \cdot (y - y_c)\right]^{-5/4} \exp\left[-\tfrac{2}{3}(y - y_c)^{3/2} (\mathrm{i}\alpha R_L u_c')^{1/2}\right]$$

The inner solutions (97) are valid provided $|y - y_c| \ll |\varepsilon|^{3/5}$. Thus, if $1 \ll |\eta| \ll |\varepsilon|^{-2/5}$, and $|\mathrm{Ph}(\eta)| < \pi$, $\chi_0^{(3)}$ becomes

$$\chi_0^{(3)}(\eta) \sim \frac{1}{2\sqrt{\pi}}\, \eta^{-5/4}\, \mathrm{e}^{-\frac{2}{3}\eta^{3/2}}$$

If these limiting forms are both valid in the overlap domain $|\varepsilon| \ll |y - y_c| \ll |\varepsilon|^{3/5}$, with $|\mathrm{Ph}(\eta)| < \pi$, their matching gives

$$C_2 = \frac{1}{2\sqrt{\pi}} (\varepsilon u_c')^{5/4}$$

(i) Heisenberg's criterion

For boundary-layer type of flows, the boundary conditions are

$$y = 0, \infty : \phi, \phi' = 0 \qquad (101)$$

Noting that the required solution is given by a linear combination of the solutions $\tilde{\phi}$ of equation (21) and $\hat{\phi}$ of equation (18), the boundary condition at $y = 0$ leads to the characteristic-value equation

$$\frac{\tilde{\phi}'(0)}{\tilde{\phi}(0)} = \frac{\hat{\phi}'(0)}{\hat{\phi}(0)} \tag{102}$$

Now, from (95), one has

$$\frac{\hat{\phi}'(y)}{\hat{\phi}(y)} = -[\mathrm{i}\alpha R_{\mathrm{L}}(u-c)]^{1/2} - \frac{5}{4}\frac{u'}{u-c} \tag{103}$$

from which

$$\frac{c}{u_0'}\frac{\hat{\phi}'(0)}{\hat{\phi}(0)} = -Z^{3/2}\,\mathrm{e}^{-\mathrm{i}\pi/4} + \frac{5}{4} \tag{104}$$

where

$$Z = \left(\frac{\alpha R_{\mathrm{L}} c^3}{u_0'^2}\right)^{1/3}, \quad u'(0) = u_0'$$

Writing

$$W = \left[1 + \frac{u_0'}{c}\frac{\tilde{\phi}(0)}{\tilde{\phi}'(0)}\right]^{-1} \tag{105}$$

equation (102) becomes

$$\frac{W-1}{W} = G(Z) \tag{106}$$

where

$$G(Z) \equiv (Z^{3/2}\,\mathrm{e}^{-\mathrm{i}\pi/4} - \tfrac{5}{4})^{-1}$$

or

$$Z^{3/2} = \left[\frac{1}{G(Z)} + \frac{5}{4}\right]\mathrm{e}^{\mathrm{i}\pi/4} \tag{107}$$

Also, note that

$$G(Z) = \begin{cases} -\dfrac{4}{5}\left(1 + \dfrac{4}{5}\mathrm{e}^{-\mathrm{i}\pi/4} Z^{3/2} + \dots\right) \\ \dfrac{\mathrm{e}^{\mathrm{i}\pi/4}}{Z^{3/2}}\left(1 + \dfrac{5}{4}\dfrac{\mathrm{e}^{\mathrm{i}\pi/4}}{Z^{3/2}} + \dots\right) \end{cases} \tag{108}$$

Recall that the inviscid solutions are

$$\tilde{\phi}_2(z) = \begin{cases} 1 + \ldots + \left[\dfrac{u_c''}{u_c'}\tilde{\phi}_1(z)\right] [\ln|z| - i\pi], & z < 0 \\ 1 + \ldots + \left[\dfrac{u_c''}{u_c'}\tilde{\phi}_1(z)\right] [\ln z], & z > 0 \end{cases} \tag{109}$$

where

$$\tilde{\phi}_1(z) = z + \frac{1}{2}\frac{u_c''}{u_c'}z^2 + \ldots, \qquad z \equiv y - y_c$$

Using (109), one finds from (106), for real values of $c = c_s$,

$$\text{Im}\,[G(Z)] = -\pi\frac{(u'')_{u=c_s}}{c_s} \tag{110}$$

Using (108), (110) gives

$$\text{Im}\,[c] = \left(\frac{u_0'^2}{\alpha_s R_\text{L}}\right)^{1/2}\left[\frac{\pi}{c_s}(u'')_{u=c_s} + \frac{5}{4}\right]\text{e}^{(i\pi)/4} \tag{111}$$

If one restricts oneself to cases where α and c do not approach zero along the neutral curve as $R_\text{L} \to \infty$, i.e. $\alpha = \alpha_s \neq 0$, $c = c_s \neq 0$, one finds from (101):

$$c_\text{i} > 0$$

or that the disturbance $\alpha = \alpha_s$, which is neutrally stable ($c = c_s$) in the inviscid case, is unstable when viscous effects appear. One therefore has Heisenberg's criterion: *If a parallel flow has an inviscid neutral disturbance with $\alpha = \alpha_s \neq 0$ and $c = c_s \neq 0$, this disturbance becomes unstable when viscous effects appear.*

Thus, for boundary-layer flows, instability prevails for large Reynolds numbers, no matter whether the inviscid instability criterion

$$(u'')_{u=c} = 0$$

is satisfied or not.

(ii) The neutral-stability curve

Since equation (18) is of the fourth order, the solution consists of a linear combination of four basic solutions:

$$\phi = C_1\phi_1 + C_2\phi_2 + C_3\phi_3 + C_4\phi_4 \tag{112}$$

where C_1, C_2, C_3, C_4 are arbitrary constants. The functions ϕ_1, ϕ_2 are the inviscid solutions, and ϕ_3, ϕ_4 are the viscous solutions. The viscous effects become important both near the critical point $u = c$ and near a rigid wall. When considering a rigid wall, one of the viscous solutions (ϕ_3) decreases exponentially with distance from the wall and is the one which is physically realistic, while the other (ϕ_4) increases exponentially with distance from the wall.

For plane Poiseuille flow, u is an even function of y, so that the general solution, ϕ, of equation (18) may be split into odd and even solutions. If one considers only the even solution, corresponding to an antisymmetric disturbance, one may focus attention on the half-channel $y = -1$ to $y = 0$. If one ignores the solution which increases exponentially with distance from the wall, one has the following three boundary conditions:

$$\begin{aligned} y = -1 &: \phi, \phi' = 0 \\ y = 0 &: \quad \phi' = 0 \end{aligned} \tag{113}$$

For a boundary-layer flow near a flat plate, one may assume that the viscous effects are negligible for the disturbance at the edge of, and outside, the boundary layer. Since the flow velocity u is constant in this region, the two inviscid solutions will be proportional to $e^{\pm \alpha y}$, of which only $e^{-\alpha y}$ is physically realistic. Thus, the three boundary conditions appropriate for boundary-layer flows are

$$\begin{aligned} y = 0 &: \phi, \phi' = 0 \\ y \to \infty &: \phi' + \alpha\phi = 0 \end{aligned} \tag{114}$$

For the Poiseuille flow, (113) results in the following characteristic-value relation:

$$\frac{\phi_{3\,\mathrm{w}}}{\phi'_{3\,\mathrm{w}}} = \frac{\phi_{1\,\mathrm{w}}\phi'_{2\,\mathrm{o}} - \phi_{2\,\mathrm{w}}\phi'_{1\,\mathrm{o}}}{\phi'_{1\,\mathrm{w}}\phi'_{2\,\mathrm{o}} - \phi'_{2\,\mathrm{w}}\phi'_{1\,\mathrm{o}}} \tag{115}$$

where the second suffix, w and o, denotes conditions at the wall or at the centre of the channel, respectively.

For the boundary-layer flow, (114) results in the following characteristic-value relation:

$$\frac{\phi_{3\,\mathrm{w}}}{\phi'_{3\,\mathrm{w}}} = \frac{\phi_{1\,\mathrm{w}}\;\Phi_{2\,\infty} - \phi_{2\,\mathrm{w}}\Phi_{1\,\infty}}{\phi'_{1\,\mathrm{w}}\;\Phi_{2\,\infty} - \phi'_{2\,\mathrm{w}}\Phi_{1\,\infty}} \tag{116}$$

where $\Phi_{n\,\infty} \equiv \phi'_{n\,\infty} + \alpha\phi_{n\,\infty}$, and the second suffix, w or ∞, denotes the wall or the edge of the boundary layer, respectively.

The right-hand sides of (115) and (116) can be calculated by using the inviscid solutions. The left-hand sides can be calculated by using $\phi_3 = \chi_0^{(3)}$. If $y = y_1$ denotes the wall ($y_1 = -1$ for plane Poiseuille flow and o for the boundary layer), (115) becomes

$$F(Y) = \frac{1}{y_1 - y_c}\left[\frac{\phi_{1\,\mathrm{w}}\phi'_{2\,\mathrm{o}} - \phi_{2\,\mathrm{w}}\phi'_{1\,\mathrm{o}}}{\phi'_{1\,\mathrm{w}}\phi'_{2\,\mathrm{o}} - \phi'_{2\,\mathrm{w}}\phi'_{1\,\mathrm{o}}}\right] \tag{117}$$

and (116) becomes

$$F(Y) = \frac{1}{y_1 - y_c}\left[\frac{\phi_{1\,\mathrm{w}}\Phi_{2\,\infty} - \phi_{2\,\mathrm{w}}\Phi_{1\,\infty}}{\phi'_{1\,\mathrm{w}}\Phi_{2\,\infty} - \phi'_{2\,\mathrm{w}}\Phi_{1\,\infty}}\right] \tag{118}$$

where

$$Y \equiv -(\eta)_{\text{wall}} = (\alpha R_L u_c')^{1/3}(y_c - y_1)$$

$$F(Y) \equiv \frac{-\int_{\infty}^{-y} \mathrm{d}\eta \int_{\infty}^{\eta} \eta^{1/2} H_{1/3}^{(1)}\left[\tfrac{2}{3}(\mathrm{i}\eta)^{3/2}\right] \mathrm{d}\eta}{Y \int_{\infty}^{y} \eta^{1/2} H_{1/3}^{(1)}\left[\tfrac{2}{3}(\mathrm{i}\eta)^{3/2}\right] \mathrm{d}\eta}$$

the function $F(Y)$ was tabulated by Tietjens (1925), Lin (1945), Holstein (1950) and Lock (1954).

The characteristic-value relation, (117) or (118), depends on α, c_r and R_L. Since it is a complex relation, it may be separated into two real relations. If c_r is eliminated, one then has a relation between α and R_L. The neutral-stability curve is the result of plotting α against R_L. The neutral-stability curves for free-shear-layer flows and the boundary-layer flows are shown in Fig I.2. The region bounded by the two asymptotic branches of the curve signifies instability. For free-shear-layer flows, the two branches of the curves approach two different asymptotes, $\alpha = 0$ and $\alpha = \alpha_s$, leaving a finite unstable region for $R_L \to \infty$ (Curle, 1956). For boundary-layer flows, the two branches approach the same asymptote $\alpha \to 0$, so that one has only the disturbance $\alpha \to 0$ at $R_L \to \infty$ (Tollmien, 1935; Schlichting, 1933; Lin, 1945). The lowest Reynolds number on each curve is the critical Reynolds number, R_{L_c}.

In general, the relation $\alpha = \alpha(R_L)$ must be obtained through computer calculations (Thomas, 1953; Shen, 1954; Orszag, 1971; Davey, 1973), although (for large Reynolds numbers) several analytical representations are known in the form of 'uniform approximations' (Lakin *et al.*, 1978) and 'triple-deck' approximations (Hall and Smith, 1982).

(5) THE NONLINEAR THEORY

According to the linear-stability calculations cited at the end of section (3), for $R_L > 5770$ (based on the maximum velocity at the centre of the channel and its half-width) there is a band of wavenumbers α, for which $c_i > 0$ – an instability. Although some experiments (Nishioka *et al.*, 1975) did verify these results, some other experiments (Davies and White, 1928) showed instability at much lower values of R_L. In order to explain this, nonlinear effects were brought into the theory.

The first paper to appear with really definite and coherent predictions about the effects of nonlinearity on instability was that of Landau (1944). Although Landau's work appeared to be conjectural in part, many of the ideas advanced there have later been justified more rigorously. In the nonlinear theory, one considers a perturbation of the form $A(t)\,\mathrm{e}^{\mathrm{i}\alpha(x - c_r t)}$ where c_r is real; $A(t)$ is found to satisfy an equation of the form (Landau, 1944; Stuart, 1958 and 1960; Watson, 1960)

$$\frac{dA}{dt} = \alpha c_i A + kA\,|A|^2 \tag{119}$$

where αc_i is the growth rate determined by the linear characteristic-value relation. Solving equation (119), one obtains

$$A^2 = \frac{\alpha C_i C\, e^{2\alpha c_i t}}{1 - kC\, e^{2\alpha c_i t}} \tag{120}$$

Equation (120) shows that:

(i) if $k < 0$, the solution $A = 0$ is unstable for $\alpha c_i > 0$ and then A approaches $(-\alpha c_i/k)^{1/2}$ as $t \to \infty$; this new steady-state solution is called a supercritical equilibrium;

(ii) if $k > 0$ and $\alpha c_i < 0$, then the solution $A = 0$ is stable for $A^2(0) < (-\alpha c_i/k)$ but is unstable for $A^2(0) > (-\alpha c_i/k)$; this instability is called a subcritical instability.

In the case of a supercritical equilibrium ($R_L > R_{L_c}$), the nonlinear effects bring an exponentially-growing disturbance to a new equilibrium state; in the case of a subcritical instability ($R_L < R_{L_c}$), the nonlinear effects produce instability if the amplitude exceeds a threshold (Meksyn and Stuart, 1951; Reynolds and Potter, 1967; Pekeris and Shkoller, 1967).

3

Governing Equations for Linearised Disturbances in Parallel Compressible Flows

INTRODUCTION

The governing equations of flow are, in the following, linearised with respect to small disturbances of the basic parallel steady flow. The method of normal modes is then used to develop linearised ordinary-differential equations governing the amplitudes of the disturbances. We will also make a search for a possible equivalence between three-dimensional disturbances and two-dimensional disturbances in the flow.

(1) LINEARISED-DISTURBANCE EQUATIONS

We consider here only the simplest model of a compressible fluid, i.e. a gas with constant specific heats C_p^*, C_v^*, constant Prandtl number $\mu^* C_p^*/k^*$ equal to unity, constant viscosity μ^*, and constant thermal conductivity k^*. One has the following governing equations:

continuity:

$$\frac{\partial \rho^*}{\partial t^*} + \frac{\partial}{\partial x^*}(\rho^* u^*) + \frac{\partial}{\partial y^*}(\rho^* v^*) + \frac{\partial}{\partial z^*}(\rho^* w^*) = 0 \qquad (1)$$

motion:

$$\frac{\partial u^*}{\partial t^*} + u^* \frac{\partial u^*}{\partial x^*} + v^* \frac{\partial u^*}{\partial y^*} + w^* \frac{\partial u^*}{\partial z^*}$$
$$= \frac{1}{\rho^*}\left(\frac{\partial \tau^*_{x^*x^*}}{\partial x^*} + \frac{\partial \tau^*_{x^*y^*}}{\partial y^*} + \frac{\partial \tau^*_{z^*y^*}}{\partial z^*}\right) \tag{2}$$

$$\frac{\partial v^*}{\partial t^*} + u^* \frac{\partial v^*}{\partial x^*} + v^* \frac{\partial v^*}{\partial y^*} + w^* \frac{\partial v^*}{\partial z^*}$$
$$= \frac{1}{\rho^*}\left(\frac{\partial \tau^*_{y^*x^*}}{\partial x^*} + \frac{\partial \tau^*_{y^*y^*}}{\partial y^*} + \frac{\partial \tau^*_{y^*z^*}}{\partial z^*}\right) \tag{3}$$

$$\frac{\partial w^*}{\partial t^*} + u^* \frac{\partial w^*}{\partial x^*} + v^* \frac{\partial w^*}{\partial y^*} + w^* \frac{\partial w^*}{\partial z^*}$$
$$= \frac{1}{\rho^*}\left(\frac{\partial \tau^*_{z^*x^*}}{\partial x^*} + \frac{\partial \tau^*_{z^*y^*}}{\partial y^*} + \frac{\partial \tau^*_{z^*z^*}}{\partial z^*}\right) \tag{4}$$

energy:

$$\rho^* C_\text{v}^* \left[\frac{\partial T^*}{\partial t^*} + u^* \frac{\partial T^*}{\partial x^*} + v^* \frac{\partial T^*}{\partial y^*} + w^* \frac{\partial T^*}{\partial z^*}\right]$$
$$= (\tau^*_{x^*x^*}\,\epsilon^*_{x^*x^*} + \tau^*_{y^*y^*}\,\epsilon^*_{y^*y^*} + \tau^*_{z^*z^*}\,\epsilon^*_{z^*z^*}$$
$$+ 2\tau^*_{x^*y^*}\,\epsilon^*_{x^*y^*} + 2\tau^*_{y^*z^*}\,\epsilon^*_{y^*z^*} + 2\tau^*_{z^*x^*}\,\epsilon^*_{z^*x^*})$$
$$+ \frac{\partial}{\partial x^*}\left(k^* \frac{\partial T^*}{\partial x^*}\right) + \frac{\partial}{\partial y^*}\left(k^* \frac{\partial T^*}{\partial y^*}\right)$$
$$+ \frac{\partial}{\partial z^*}\left(k^* \frac{\partial T^*}{\partial z^*}\right) \tag{5}$$

state:

$$p^* = \rho^* R^* T^* \tag{6}$$

where

$$\epsilon^*_{x^*x^*} = \frac{\partial u^*}{\partial x^*}, \quad \epsilon^*_{y^*y^*} = \frac{\partial v^*}{\partial y^*}, \quad \epsilon^*_{z^*z^*} = \frac{\partial w^*}{\partial z^*}$$

$$\epsilon^*_{x^*y^*} = \frac{1}{2}\left(\frac{\partial u^*}{\partial y^*} + \frac{\partial v^*}{\partial x^*}\right), \quad \epsilon^*_{y^*z^*} = \frac{1}{2}\left(\frac{\partial v^*}{\partial z^*} + \frac{\partial w^*}{\partial y^*}\right)$$

$$
\begin{aligned}
\epsilon^*_{z^*x^*} &= \frac{1}{2}\left(\frac{\partial w^*}{\partial x^*} + \frac{\partial u^*}{\partial z^*}\right) \\
\tau^*_{x^*x^*} &= 2\mu^*\,\epsilon^*_{x^*x^*} - \tfrac{2}{3}\mu^*\,(\epsilon^*_{x^*x^*} + \epsilon^*_{y^*y^*} + \epsilon^*_{z^*z^*}) - p^* \\
\tau^*_{y^*y^*} &= 2\mu^*\,\epsilon^*_{y^*y^*} - \tfrac{2}{3}\mu^*\,(\epsilon^*_{x^*x^*} + \epsilon^*_{y^*y^*} + \epsilon^*_{z^*z^*}) - p^* \\
\tau^*_{z^*z^*} &= 2\mu^*\,\epsilon^*_{z^*z^*} - \tfrac{2}{3}\mu^*\,(\epsilon^*_{x^*x^*} + \epsilon^*_{y^*y^*} + \epsilon^*_{z^*z^*}) - p^* \\
\tau^*_{x^*y^*} &= 2\mu^*\,\epsilon^*_{x^*y^*}, \qquad \tau^*_{y^*z^*} = 2\mu^*\,\epsilon^*_{y^*z^*} \\
\tau^*_{z^*x^*} &= 2\mu^*\,\epsilon^*_{z^*x^*}
\end{aligned}
$$

(u^*, v^*, w^*) are the velocity components in the x^*, y^*, z^* directions, p^* the pressure, ρ^* the density, T^* the temperature, and the asterisks refer to dimensional quantities.

Consider now a motion which is slightly perturbed from a two-dimensional steady state. Divide all variables into a mean-flow part and a small-disturbance part, i.e.

$$Q^*(x^*, y^*, z^*; t^*) = \bar{Q}^*(x^*, y^*) + Q^{*\prime}(x^*, y^*, z^*; t^*) \tag{7}$$

where x^* is the distance along the flow, and y^* and z^* perpendicular to the flow.

For the case of a homogeneous incompressible fluid as we saw in Chapter 2, Squire (1933) showed by means of a transformation of the equations of flow that to every amplified three-dimensional disturbance corresponds a faster-growing two-dimensional disturbance. This greatly simplified the analysis for the case of an incompressible fluid. For the case of a compressible fluid, one may attempt such a transformation, but one cannot establish the mathematical equivalence between the boundary-value problems associated with a three-dimensional disturbance and a two-dimensional disturbance (see Appendix). Besides, the latter problem (even if its mathematical equivalence is established) does not represent a proper two-dimensional disturbance. Moreover, the interpretation of the final results, unlike that in the case of an incompressible fluid, becomes very difficult, if not impossible. For instance, Squire's conclusion in the case of an incompressible fluid can be very simply derived by a rotation of axes, as indicated by Lin (1953). But for the case of a compressible fluid, a vector quantity, such as velocity, and scalar quantity, such as temperature, do not behave in the same manner during such a transformation. As Lin (1953) pointed out, the temperature distribution which is preserved in the rotation of the axes, depends on the actual free-stream velocity, and not merely its component along the direction of the wave propagation. Therefore, the temperature and velocity distributions in the equivalent two-dimensional problem do not correspond to those of another physically-possible parallel flow.

However, notwithstanding the absence of a definitive justification, we shall consider in the following only the two-dimensional disturbances purely for the sake of a considerable mathematical simplicity that follows.

Thus, let

$$Q^*(x^*, y^*, z^*; t^*) = \bar{Q}^*(x^*, y^*) + Q^{*\prime}(x^*, y^*; t^*) \tag{8}$$

Decompose all the variables in equations (1)–(6) according equation (8), noting that the steady-state parts satisfy equations (1)–(6) by themselves, and neglecting terms quadratic in the small disturbance, there follows the system of linearised equations for the disturbances:

$$\frac{\partial \rho^{*\prime}}{\partial t^*} + \frac{\partial}{\partial x^*}(\rho^{*\prime}\,\bar{u}^* + \bar{\rho}^*\,u^{*\prime}) + \frac{\partial}{\partial y^*}(\rho^{*\prime}\,\bar{v}^* + \bar{\rho}^*\,v^{*\prime}) = 0 \tag{9}$$

$$\begin{aligned}\frac{\partial u^{*\prime}}{\partial t^*} &+ \left(u^{*\prime}\,\frac{\partial \bar{u}^*}{\partial x^*} + v^{*\prime}\,\frac{\partial \bar{u}^*}{\partial y^*}\right) + \left(\bar{u}^*\,\frac{\partial u^{*\prime}}{\partial x^*} + \bar{v}^*\,\frac{\partial u^{*\prime}}{\partial y^*}\right)\\ &= \frac{1}{\bar{\rho}^*}\left(\frac{\partial \tau^{*\prime}_{x^*x^*}}{\partial x^*} + \frac{\partial \tau^{*\prime}_{x^*y^*}}{\partial y^*}\right)\\ &\quad - \frac{\rho^{*\prime}}{\bar{\rho}^{*2}}\left(\frac{\partial \bar{\tau}^*_{x^*x^*}}{\partial x^*} + \frac{\partial \bar{\tau}^*_{x^*y^*}}{\partial y^*}\right)\end{aligned} \tag{10}$$

$$\begin{aligned}\frac{\partial v^{*\prime}}{\partial t^*} &+ \left(u^{*\prime}\,\frac{\partial \bar{v}^*}{\partial x^*} + v^{*\prime}\,\frac{\partial \bar{v}^*}{\partial y^*}\right) + \left(\bar{u}^*\,\frac{\partial v^{*\prime}}{\partial x^*} + \bar{v}^*\,\frac{\partial v^{*\prime}}{\partial y^*}\right)\\ &= \frac{1}{\bar{\rho}^*}\left(\frac{\partial \tau^{*\prime}_{x^*y^*}}{\partial x^*} + \frac{\partial \tau^{*\prime}_{y^*y^*}}{\partial y^*}\right)\\ &\quad - \frac{\rho^{*\prime}}{\bar{\rho}^{*2}}\left(\frac{\partial \bar{\tau}^*_{x^*y^*}}{\partial x^*} + \frac{\partial \bar{\tau}^*_{y^*y^*}}{\partial y^*}\right)\end{aligned} \tag{11}$$

$$\begin{aligned}C_V^*\,\bar{\rho}^*&\left(\frac{\partial T^{*\prime}}{\partial t^*} + \bar{u}^*\,\frac{\partial T^{*\prime}}{\partial x^*} + \bar{v}^*\,\frac{\partial T^{*\prime}}{\partial y^*} + u^{*\prime}\,\frac{\partial \bar{T}^*}{\partial x^*} + v^{*\prime}\,\frac{\partial \bar{T}^*}{\partial y^*}\right)\\ &+ C_V^*\,\rho^{*\prime}\left(\bar{u}^*\,\frac{\partial \bar{T}^*}{\partial x^*} + \bar{v}^*\,\frac{\partial \bar{T}^*}{\partial y^*}\right)\\ = &\,\frac{\partial}{\partial x^*}\left(k^*\,\frac{\partial T^{*\prime}}{\partial x^*}\right) + \frac{\partial}{\partial y^*}\left(k^*\,\frac{\partial T^{*\prime}}{\partial y^*}\right)\\ &+ (\bar{\tau}^*_{x^*x^*}\,\epsilon^{*\prime}_{x^*x^*} + 2\bar{\tau}^*_{x^*y^*}\,\epsilon^{*\prime}_{x^*y^*} + \bar{\tau}^*_{y^*y^*}\,\epsilon^{*\prime}_{y^*y^*})\\ &+ (\tau^{*\prime}_{x^*x^*}\,\bar{\epsilon}^*_{x^*x^*} + 2\tau^{*\prime}_{x^*y^*}\,\bar{\epsilon}^*_{x^*y^*} + \tau^{*\prime}_{y^*y^*}\,\bar{\epsilon}^*_{y^*y^*})\end{aligned} \tag{12}$$

$$\frac{p^{*\prime}}{\bar{p}^*} = \frac{\rho^{*\prime}}{\bar{\rho}^*} + \frac{T^{*\prime}}{\bar{T}^*} \qquad (13)$$

Next, nondimensionalise the variables with respect to the reference quantities:

$$\bar{V}_1^*,\ \ \bar{p}_1^*,\ \ \bar{\rho}_1^*,\ \ \bar{T}_1^*,\ \ \mu^*,\ \ k^*,\ \ C_V^*,\ \ L^*$$

where the subscript '1' on the first four quantities corresponds to the conditions in the free stream for the case of a boundary layer.

Note also that a minor difficulty is encountered in the above because there is no natural length for the x-direction in the problem to which the relevant spatial coordinates could be referred. We have circumvented this by selecting an arbitrary reference length L^*.

We are restricting our study to parallel flows so that we shall neglect

(i) terms involving mean normal velocity $\bar{v}^*$,
(ii) longitudinal derivatives of the mean-flow quantities as compared with their normal derivatives,
i.e.

$$\bar{u}^* \gg \bar{v}^*$$

$$\frac{\partial \bar{Q}^*}{\partial x^*} \ll \frac{\partial \bar{Q}^*}{\partial y^*} \qquad (14)$$

Note that this approximation is, however, not valid for the disturbances. (Saric and Nayfeh (1975) made a study of the effects of departures from parallel flows and found that they become prominent in situations involving adverse pressure gradients and result in a drop in the minimum critical Reynolds number.)

Then on obtains from equations (9)–(13)

$$\frac{\partial \rho'}{\partial t} + \bar{\rho}\left(\frac{\partial u'}{\partial x} + \frac{\partial v'}{\partial y}\right) + v'\frac{\partial \bar{\rho}}{\partial y} + \bar{u}\frac{\partial \rho'}{\partial x} = 0 \qquad (15)$$

$$\bar{\rho}\left(\frac{\partial u'}{\partial t} + \bar{u}\frac{\partial u'}{\partial x} + v'\frac{\partial \bar{u}}{\partial y}\right) = -\frac{1}{\sigma M_1^2}\frac{\partial p'}{\partial x} +$$

$$+ \frac{1}{R_L}\left[2\frac{\partial^2 u'}{\partial x^2} + \frac{\partial^2 u'}{\partial y^2} + \frac{\partial^2 v'}{\partial x \partial y} + \right.$$

$$\left. - \frac{2}{3}\left(\frac{\partial^2 u'}{\partial x^2} + \frac{\partial^2 v'}{\partial x \partial y}\right)\right] \qquad (16)$$

$$\bar{\rho}\left(\frac{\partial v'}{\partial t}+\bar{u}\frac{\partial v'}{\partial x}\right)=-\frac{1}{\sigma M_1^2}\frac{\partial p'}{\partial y}+$$

$$+\frac{1}{R_L}\left[2\frac{\partial^2 v'}{\partial y^2}+\frac{\partial^2 v'}{\partial x^2}+\frac{\partial^2 u'}{\partial x\partial y}+\right.$$

$$\left.-\frac{2}{3}\left(\frac{\partial^2 u'}{\partial x\partial y}+\frac{\partial^2 v'}{\partial y^2}\right)\right] \tag{17}$$

$$\bar{\rho}\left(\frac{\partial T'}{\partial t}+\bar{u}\frac{\partial T'}{\partial x}+v'\frac{\partial \bar{T}}{\partial y}\right)$$

$$=-\bar{\rho}\bar{T}(\sigma-1)\left(\frac{\partial u'}{\partial x}+\frac{\partial v'}{\partial y}\right)+$$

$$+\frac{\sigma}{R_L}\left(\frac{\partial^2 T'}{\partial x^2}+\frac{\partial^2 T'}{\partial y^2}\right)+$$

$$+\sigma(\sigma-1)\frac{M_1^2}{R_L}\left[2\frac{\partial \bar{u}}{\partial y}\left(\frac{\partial u'}{\partial y}+\frac{\partial v'}{\partial x}\right)\right] \tag{18}$$

$$p'=\bar{\rho}T'+\rho'\bar{T} \tag{19}$$

where the nondimensional parameters are

$$\sigma=\frac{C_p^*}{C_v^*},\quad M_1^2=\frac{\bar{V}_1^{*2}}{\sigma^* R^* \bar{T}_1^*},\quad R_L=\frac{\bar{\rho}_1^*\bar{V}_1^* L^*}{\mu^*}$$

The stability problem involves the growth of an arbitrary infinitesimal perturbation, and in the theory of normal modes we adopt here, such perturbations are resolved into independent wave-like components. Each component is supposed to satisfy the linearised equations (15)–(19) and the associated boundary conditions (which we shall discuss soon) separately. Moreover, that the coefficients in equations (15)–(19) are independent of x suggests the consideration of disturbances that are spatially periodic in x. Since the variables x, t occur in equations (15)–(19) only in the form of partial derivatives, and are cyclic, the system (15)–(19) admits solutions which are exponential functions of x, t, with amplitudes that depend only on y. If the solutions are to be bounded for x becoming both $+\infty$ and $-\infty$, the corresponding exponent must be purely imaginary. Thus one assumes the disturbances of the form

$$Q'(x,y;t)=q(y)\,\mathrm{e}^{\mathrm{i}\alpha(x-ct)} \tag{20}$$

for a typical wave component travelling in the x-direction. Note that α is real and $2\pi/\alpha = \lambda$ is the wavelength of the disturbance, and c is the complex wave velocity:

$$c = c_{\mathrm{r}} + \mathrm{i}c_{\mathrm{i}}$$

It is to be understood that the real parts are to be taken to obtain physical quantities, this being permissible for the linear problem. Each component travels as a wave in the direction of $(\alpha, 0)$ with phase speed αc_{r} and grows or dies away in time like $\mathrm{e}^{\alpha c_{\mathrm{i}} t}$. Thus if $\alpha c_{\mathrm{i}} > 0$, the disturbance is amplified in time, and the flow is unstable, whereas if $\alpha c_{\mathrm{i}} < 0$, the disturbance is attenuated, and the flow is stable. The flow is said to be neutrally stable if $\alpha c_{\mathrm{i}} = 0$. (Considerations of disturbances that grow or decay in space instead of in time have also been made in the literature, e.g. by Gaster (1962).)

The method of normal modes, which represents a disturbance by wave components, thus serves to separate the variables and reduces the linearised equations (15)–(19) from partial-differential equations to ordinary-differential equations. If the complex amplitude functions of the flow disturbance are defined by

$$(u', v', p', \rho', T') = (f, \alpha\varphi, \pi, \gamma, \theta)\, \mathrm{e}^{\mathrm{i}\alpha(x - ct)} \tag{21}$$

equations (15)–(19) give for the linearised amplitudes of small disturbances in a parallel flow:

$$\mathrm{i}(u - c)\gamma + \rho(\varphi' + \mathrm{i}f) + \rho'\varphi = 0 \tag{22}$$

$$\alpha\rho\,[\mathrm{i}(u - c)f + u'\varphi] = -\frac{\mathrm{i}\alpha\pi}{\sigma M_1^2} + \frac{1}{R_{\mathrm{L}}}\,[f'' + \alpha^2\,(\mathrm{i}\varphi' - 2f)] + -\frac{2}{3}\frac{\alpha^2}{R_{\mathrm{L}}}\,(-f + \mathrm{i}\varphi') \tag{23}$$

$$\alpha^2\rho\,[\mathrm{i}(u - c)\varphi] = -\frac{\pi'}{\sigma M_1^2} + \frac{\alpha}{R_{\mathrm{L}}}\,(2\varphi'' + \mathrm{i}f' - \alpha^2\varphi) + -\frac{2}{3}\frac{\alpha}{R_{\mathrm{L}}}\,(-\varphi'' + \mathrm{i}f') \tag{24}$$

$$\alpha\rho\,[\mathrm{i}(u - c)\theta + T'\varphi] = -\alpha\rho T(\sigma - 1)\,(\varphi' + \mathrm{i}f) + + \frac{\alpha}{R_{\mathrm{L}}}\,(\theta'' - \alpha^2\theta) + \frac{\sigma(\sigma - 1)M_1^2}{R_{\mathrm{L}}}\,[2u'(f' + \mathrm{i}\alpha^2\varphi)] \tag{25}$$

$$\frac{\pi}{p} = \frac{\gamma}{\rho} + \frac{\theta}{T} \tag{26}$$

In equations (22)–(26), the bars overhead have been dropped, and primes denote differentiation with respect to y.

(2) BOUNDARY CONDITIONS

Usually, all disturbance amplitudes vanish at infinity so that

$$y \to \infty; \quad f, \varphi, \pi, \gamma, \theta \to 0 \tag{27}$$

The other boundary conditions are those of nonslip conditions at a solid boundary and the stress conditions at a free surface or an interface. In the stability problem, the boundary conditions are thus homogeneous. We shall henceforth take $\alpha \geqslant 0$ without loss of generality. Then the criterion for instability of the basic flow is that there be a solution with $c_i > 0$ for some $\alpha > 0$.

These boundary conditions lead to a characteristic-value problem of the form

$$c = F(\alpha, R_L, M_1^2) \tag{28}$$

APPENDIX: SEARCH FOR AN EQUIVALENCE BETWEEN THREE-DIMENSIONAL DISTURBANCES AND TWO-DIMENSIONAL DISTURBANCES

The linearised equations for a three-dimensional disturbance in a parallel flow of compressible fluid are

$$\frac{\partial \rho'}{\partial t} + \bar{\rho}\left(\frac{\partial u'}{\partial x} + \frac{\partial v'}{\partial y} + \frac{\partial w'}{\partial z}\right) + v'\,\frac{\partial \bar{\rho}}{\partial y} + \bar{u}\,\frac{\partial \rho'}{\partial x} = 0 \tag{A.1}$$

$$\begin{aligned}\bar{\rho}\left(\frac{\partial u'}{\partial t} + \bar{u}\,\frac{\partial u'}{\partial x} + v'\,\frac{\partial \bar{u}}{\partial y}\right) = &-\frac{1}{\sigma M_1^2}\,\frac{\partial p'}{\partial x} + \\ &+\frac{1}{R_L}\left[2\,\frac{\partial^2 u'}{\partial x^2} + \frac{\partial^2 u'}{\partial y^2} + \frac{\partial^2 v'}{\partial x \partial y} + \frac{\partial^2 u'}{\partial z^2} + \right.\\ &\left. + \frac{\partial^2 w'}{\partial x \partial z} - \frac{2}{3}\left(\frac{\partial^2 u'}{\partial x^2} + \frac{\partial^2 v'}{\partial x \partial y} + \frac{\partial^2 w'}{\partial x \partial z}\right)\right]\end{aligned} \tag{A.2}$$

$$\begin{aligned}\bar{\rho}\left(\frac{\partial v'}{\partial t} + \bar{u}\,\frac{\partial v'}{\partial x}\right) = &-\frac{1}{\sigma M_1^2}\,\frac{\partial p'}{\partial y} + \\ &+\frac{1}{R_L}\left[2\,\frac{\partial^2 v'}{\partial y^2} + \frac{\partial^2 v'}{\partial x^2} + \frac{\partial^2 u'}{\partial x \partial y} + \frac{\partial^2 v'}{\partial z^2} + \right.\\ &\left. + \frac{\partial^2 w'}{\partial y \partial z} - \frac{2}{3}\left(\frac{\partial^2 v'}{\partial y^2} + \frac{\partial^2 u'}{\partial x \partial y} + \frac{\partial^2 w'}{\partial y \partial z}\right)\right]\end{aligned} \tag{A.3}$$

$$\bar{\rho}\left(\frac{\partial w'}{\partial t}+\bar{u}\,\frac{\partial w'}{\partial x}\right)=-\,\frac{1}{\sigma M_1^2}\,\frac{\partial p'}{\partial z}+$$

$$+\,\frac{1}{R_L}\left[2\,\frac{\partial^2 w'}{\partial z^2}+\frac{\partial^2 w'}{\partial x^2}+\frac{\partial^2 u'}{\partial x\partial z}+\frac{\partial^2 v'}{\partial y\partial z}+\right.$$

$$\left.+\,\frac{\partial^2 w'}{\partial y^2}-\frac{2}{3}\left(\frac{\partial^2 w'}{\partial z^2}+\frac{\partial^2 u'}{\partial x\partial z}+\frac{\partial^2 v'}{\partial y\partial z}\right)\right] \tag{A.4}$$

$$\bar{\rho}\left(\frac{\partial T'}{\partial t}+\bar{u}\,\frac{\partial T'}{\partial x}+v'\,\frac{\partial \bar{T}}{\partial y}\right)=-\,\rho T(\sigma-1)\left(\frac{\partial u'}{\partial x}+\frac{\partial v'}{\partial y}+\frac{\partial w'}{\partial z}\right)$$

$$+\,\frac{\sigma}{R_{\rm L}}\left(\frac{\partial^2 T'}{\partial x^2}+\frac{\partial^2 T'}{\partial y^2}+\frac{\partial^2 T'}{\partial z^2}\right)+$$

$$+\,\sigma(\sigma-1)\,\frac{M_1^2}{R_{\rm L}}\left[2\,\frac{\partial\bar{u}}{\partial y}\left(\frac{\partial u'}{\partial y}+\frac{\partial v'}{\partial x}\right)\right] \tag{A.5}$$

$$p'=\bar{\rho}T'+\rho'\bar{T} \tag{A.6}$$

Considering decomposition into normal modes as

$$Q'(x,y,z;t)=q(y)\,\mathrm{e}^{\mathrm{i}(\alpha x+\beta z-\alpha c t)} \tag{A.7}$$

one obtains from equations (A.1)–(A.6)

$$\mathrm{i}\alpha(u-c)\gamma+\rho'\alpha\varphi+\rho(\mathrm{i}\alpha f+\alpha\varphi'+\mathrm{i}\beta\psi)=0 \tag{A.8}$$

$$\rho\,[\mathrm{i}\alpha(u-c)f+\alpha u'\varphi]=-\,\frac{\mathrm{i}\alpha\pi}{\sigma M_1^2}+$$

$$+\,\frac{1}{R_{\rm L}}\left[f''-\left(\frac{4\alpha^2}{3}+\beta^2\right)f+\frac{\mathrm{i}\alpha^2\varphi'}{3}-\frac{\alpha\beta\psi}{3}\right] \tag{A.9}$$

$$\rho\,[\mathrm{i}\alpha^2(u-c)\varphi]=-\,\frac{\pi'}{\sigma M_1^2}+$$

$$+\,\frac{1}{R_{\rm L}}\left[\frac{4\alpha}{3}\varphi''-(\alpha^2+\beta^2)\varphi+\frac{\mathrm{i}\alpha f'}{3}+\frac{\mathrm{i}\beta\psi}{3}\right] \tag{A.10}$$

$$\rho\,[\mathrm{i}\alpha(u-c)\psi]=-\,\frac{\mathrm{i}\beta\pi}{\sigma M_1^2}+$$

$$+\,\frac{1}{R_{\rm L}}\left[\psi''-\left(\alpha^2+\frac{4\beta^2}{3}\right)\psi+\frac{\mathrm{i}\beta\alpha\varphi'}{3}-\frac{\alpha\beta}{3}f\right] \tag{A.11}$$

$$\rho\,[\mathrm{i}\alpha(u-c)\theta + T'\varphi] = -\rho T(\sigma-1)\,(\mathrm{i}\alpha f + \alpha\varphi' + \mathrm{i}\beta\psi) + \\ + \frac{\sigma}{R_{\mathrm{L}}}\,(\theta'' - \alpha^2\theta - \beta^2\theta) + \\ + \sigma(\sigma-1)\,\frac{M_1^2}{R_{\mathrm{L}}}\,[2u'(f' + \mathrm{i}\alpha^2\varphi)] \tag{A.12}$$

$$\frac{\pi}{p} = \frac{\gamma}{\rho} + \frac{\theta}{T} \tag{A.13}$$

where

$$w'(x,y,z;t) = \psi(y)\,\mathrm{e}^{\mathrm{i}(\alpha x + \beta z - \alpha c t)}$$

If one makes the following transformation to that of an equivalent two-dimensional problem (if one exists?):

$$\alpha f + \beta\psi = \hat{\alpha}\hat{f}$$

$$\alpha\varphi = \hat{\alpha}\hat{\varphi}, \qquad \alpha\pi = \hat{\alpha}\hat{\pi}, \quad \alpha\gamma = \hat{\alpha}\hat{\gamma}$$

$$\alpha\theta = \hat{\alpha}\hat{\theta}, \qquad c = \hat{c}, \qquad \alpha R_{\mathrm{L}} = \hat{\alpha}\hat{R}_{\mathrm{L}}$$

$$\alpha M_1^2 = \hat{\alpha} M_1^2, \qquad \alpha^2 + \beta^2 = \hat{\alpha}^2 \tag{A.14}$$

then one obtains from equations (A.8)–(A.13)

$$\mathrm{i}\hat{\alpha}(u-c)\hat{\gamma} + \rho'\hat{\alpha}\hat{\varphi} + \rho(\mathrm{i}\hat{\alpha}\hat{f} + \hat{\alpha}\hat{\varphi}') = 0 \tag{A.15}$$

$$\rho\,[\mathrm{i}\hat{\alpha}(u-c)\hat{f} + u'\hat{\alpha}\hat{\varphi}] = -\frac{\mathrm{i}\hat{\alpha}\hat{\pi}}{\sigma M_1^2} + \\ + \frac{1}{\hat{R}_{\mathrm{L}}}\left(\hat{f}'' - \frac{4\hat{\alpha}^2}{3}\hat{f} + \frac{\mathrm{i}\hat{\alpha}}{3}\hat{\varphi}\right) \tag{A.16}$$

$$\rho\,[\mathrm{i}\hat{\alpha}^2(u-c)\hat{\varphi}] = -\frac{\hat{\pi}'}{\sigma M_1^2} + \\ + \frac{1}{\hat{R}_{\mathrm{L}}}\left(\frac{4\hat{\alpha}}{3}\hat{\varphi}'' - \hat{\alpha}^2\hat{\varphi} + \frac{\mathrm{i}\hat{\alpha}^2\hat{f}}{3}\right) \tag{A.17}$$

$$\rho\,[\mathrm{i}\hat{\alpha}(u-c)\hat{\theta} + T'\hat{\alpha}\hat{\varphi}] = -\rho T(\sigma-1)\,(\mathrm{i}\hat{\alpha}\hat{f} + \hat{\alpha}\hat{\varphi}') + \\ + \frac{\sigma}{\hat{R}_{\mathrm{L}}}\,(\hat{\theta}'' - \hat{\alpha}^2\hat{\theta}) + \sigma(\sigma-1)\,\frac{M_1^2}{R_{\mathrm{L}}}\,[2u'(f' + \mathrm{i}\alpha^2\varphi)] \tag{A.18}$$

$$\frac{\hat{\pi}}{p} = \frac{\hat{\gamma}}{\rho} + \frac{\hat{\theta}}{T} \tag{A.19}$$

Note the last term, viz.

$$\sigma(\sigma - 1)\ \frac{M_1^2}{R_L}\ [2u'(f' + i\alpha^2\varphi)]$$

in equation (A.18), which represents the heat generated due to viscous effects on the disturbances, is the one that precludes the mathematical equivalence of the problem of a three-dimensional disturbance to that of a two-dimensional disturbance.

4

Linearised Inviscid Disturbances

INTRODUCTION

We now begin a study of the equations, derived in the previous chapter, for the linearised disturbances in the flow in the inviscid limit $\alpha R_L \to \infty$. First we shall construct various kinds of solution for the inviscid disturbances, and then discuss their validity, uniqueness, regularity, and behaviour in various domains. Next, we shall develop a representation for the far-field behaviour of the inviscid disturbances using the method of multiple scales.

(1) INVISCID PROBLEM

In the limit $\alpha R_L \to \infty$, equations (III.22)–(III.27) become

$$\mathrm{i}(u-c)\gamma + \rho(\varphi' + \mathrm{i}f) + \rho'\varphi = 0 \tag{1}$$

$$\rho\,[\mathrm{i}(u-c)f + u'\varphi] = -\frac{\mathrm{i}\pi}{\sigma M_1^2} \tag{2}$$

$$\alpha^2\rho\,[\mathrm{i}(u-c)\varphi] = -\frac{\pi'}{\sigma M_1^2} \tag{3}$$

$$\rho\,[\mathrm{i}(u-c)\theta + T'\varphi] = -\rho T(\sigma-1)(\varphi' + \mathrm{i}f) \tag{4}$$

$$\frac{\pi}{p} = \frac{\gamma}{\rho} + \frac{\theta}{T} \tag{5}$$

Note that in the inviscid limit the orders of the differential equations have been reduced. This is consistent with the fact that the solutions of equations (III.22)–(III.26) have an essential singularity at $R_L \to \infty$.

Using equations (1) and (4), equation (5) gives

$$\frac{\pi}{p} = \frac{i[\sigma(\varphi' + if)]}{u - c} \tag{6}$$

Using (6), equation (2) gives

$$if = \frac{p\sigma\varphi' - \sigma\rho M_1^2 u'(u-c)\varphi}{\rho\sigma M_1^2 (u-c)^2 - p\sigma} \tag{7}$$

Using (6) and (7), equation (3) gives

$$\rho\,[i\alpha^2 (u-c)\varphi] = -\frac{p}{\sigma M_1^2}\frac{d}{dy}\left[\frac{i\sigma\varphi'}{u-c} + \right.$$

$$\left. + \frac{i\sigma}{u-c}\,\frac{p\sigma\varphi' - \sigma\rho M_1^2 u'(u-c)\varphi}{\rho\sigma M_1^2 (u-c)^2 - p\sigma}\right] \tag{8}$$

Using equation (III.6), (8) becomes

$$\frac{\alpha^2 (u-c)\varphi}{T} = -\frac{1}{\sigma M_1^2}\frac{d}{dy}\left[\frac{\sigma\varphi'}{u-c} + \right.$$

$$\left. + \frac{\sigma T\varphi' - \sigma M_1^2 u'(u-c)\varphi}{(u-c)\{M_1^2 (u-c)^2 - T\}}\right]$$

which can be written as

$$\frac{d}{dy}\left[\frac{\varphi'}{T - M_1^2 (u-c)^2}\right] - \left[\frac{\dfrac{d}{dy}\left\{\dfrac{u'}{T - M_1^2 (u-c)^2}\right\}}{u-c} + \frac{\alpha^2}{T}\right]\varphi = 0 \tag{9}$$

If the flow is bounded by two planes $y = y_1, y_2$ parallel to the flow, the boundary conditions are

$$y = y_1, y_2 : \varphi = 0 \tag{10}$$

Note that unless

$$\left[\frac{d}{dy}\left(\frac{u'}{T}\right)\right]_{u=c} = 0$$

the point $y = y_c$, where $u = c$, in the complex y-plane is a regular singularity of equation (9). Otherwise, the inviscid system (9) and (10) would have a well-defined characteristic-value problem. Nevertheless, for complete linearised-disturbance equations (III.22)–(III.26) including the viscous terms, this point is merely an ordinary point. This singularity is introduced purely by the process of taking the asymptotic limit.

The singular nature of equation (9) is also one of the chief difficulties in constructing uniformly-valid asymptotic solutions to the full equations. In particular, under certain conditions, it turns out that a regular solution of equation (9) satisfying the boundary conditions (10) does not exist if the critical point y_c lies in $y_1 < y < y_2$.

In the free stream, equation (9) becomes

$$\varphi'' - \frac{\alpha^2}{T_1}[T_1 - M_1^2(V_1 - c)^2]\varphi = 0 \tag{11}$$

from which

$$y \to \infty : \varphi = A\,e^{-s_1 y} \tag{12}$$

where

$$s_1 = \frac{\alpha}{\sqrt{T_1}}\sqrt{T_1 - M_1^2(V_1 - c)^2}$$

and A is an arbitrary constant. In order to define s_1 uniquely it is necessary to introduce a 'cut' along the negative real axis of the complex s_1-plane. Note that regardless of whether

$$[T_1 - M_1^2(V_1 - c)^2 \gtrless 0$$

the real part of s_1 will be positive as long as

$$-\frac{\pi}{2} < \arg\sqrt{T_1 - M_1^2(V_1 - c)^2} < \frac{\pi}{2}$$

(2) INVISCID SOLUTIONS

We now come to the issue of constructing solutions of equation (9). Since the coefficients of equation (9) are entire functions of the parameter α^2, two particular integrals $\varphi_1^{(1)}, \varphi_2^{(1)}$ of this equation must also be entire functions of α^2. Series developments of $\varphi_1^{(1)}, \varphi_2^{(1)}$ in powers of α^2 – an idea originally due to Heisenberg (1924) – are therefore uniformly valud for any finite region of the complex y-plane, for a fixed value of y, except when y is the singular point y_c of equation (9). Consider the special case, $\alpha^2 = 0$, first. It can be immediately verified that a solution of equation (9) in this case is

$$\varphi_{10}^{(1)} = u - c \tag{13}$$

A second solution is

$$\varphi_{20}^{(1)} = (u - c)\int_{y_1}^{y}\left[\frac{T}{(u-c)^2} - M_1^2\right]dy \tag{14}$$

One then seeks solutions in ascending powers of α^2 as

$$\varphi_1^{(1)} = \varphi_{10}^{(1)} + \alpha^2 \varphi_{11}^{(1)} + \alpha^4 \varphi_{12}^{(1)} + \ldots$$
$$\varphi_2^{(1)} = \varphi_{20}^{(1)} + \alpha^2 \varphi_{21}^{(1)} + \alpha^4 \varphi_{22}^{(1)} + \ldots \tag{15}$$

Substituting the series (12) into equation (6), and equating the coefficients of equal power of α, one obtains

$$\left[\frac{\varphi_{11}^{(1)\prime}}{T - M_1^2(u-c)^2}\right]' - \left[\frac{\left\{\dfrac{u'}{T - M_1^2(u-c)^2}\right\}'}{u-c}\right]\varphi_{11}^{(1)} = \frac{\varphi_{10}^{(1)}}{T}$$

or

$$\left[\frac{1}{T - M_1^2(u-c)^2}\left\{\frac{\varphi_{11}^{(1)}}{u-c}\right\}'(u-c)^2\right]' = \frac{(u-c)}{T}\varphi_{10}^{(1)}$$

from which

$$\varphi_{11}^{(1)} = (u-c)\int_{y_1}^{y}\left[\frac{T}{(u-c)^2} - M_1^2\right]dy\int_{y_1}^{y}\frac{(u-c)}{T}\varphi_{10}^{(1)}\,dy \tag{16}$$

In general,

$$\varphi_{1,n+1}^{(1)} = (u-c)\int_{y_1}^{y}\left[\frac{T}{(u-c)^2} - M_1^2\right]dy\int_{y_1}^{y}\frac{(u-c)}{T}\varphi_{1n}^{(1)}\,dy \tag{17}$$

Similarly,

$$\varphi_{2,n+1}^{(1)} = (u-c)\int_{y_1}^{y}\left[\frac{T}{(u-c)^2} - M_1^2\right]dy\int_{y_1}^{y}\frac{(u-c)}{T}\varphi_{2n}^{(1)}\,dy \tag{18}$$

Also, in order that $\varphi_1^{(1)}$, $\varphi_2^{(1)}$ may be valid approximations to the regular solution of equations (III.22)–(III.26), all along the path of integration between the point $y = y_1$, and $y = y_2$ on the real axis, that path must lie wholly in a region in which the asymptotic expansions of the regular solutions for large values of R_L are valid. But within the framework of the inviscid problem alone, it is not obvious in the normal-mode approach how one should negotiate the point $y = y_c$, particularly when c is real.

In analogy with the arguments made by Lin (1955) for the case of an incompressible fluid, one may try to resolve this ambiguity by arguing that the inviscid equation (9) has meaning only as the limiting case of the complete viscous equations (III.22)–(III.26), and that the inviscid solution must be a valid asymptotic representation of the complete viscous equations (III.22)–(III.26) all along the path of integration. By making a comparison with the asymptotic forms of the complete viscous solutions, one may assert (see Chapter VIII) that if one continues the path of integration around the critical point $y = y_c$ from below, one is then always in a region in the complex y-plane where the inviscid solutions are valid asymptotic representations of the complete viscous solutions.

But it turns out that (see Chapter V), if one poses the problem of stability as an initial-value problem, one will be able to resolve these difficulties without having to adopt an involved procedure of going outside the inviscid model for a gas. Besides, the initial-value-problem analysis shows that the singularity at $u = c$ leads to a continuous spectrum of characteristic values that is in addition to the discrete spectrum of characteristic values, which may be real or complex.

In order to obtain two other linearly-independent Tollmien-forms of solution $\varphi_1^{(2)}$, $\varphi_2^{(2)}$ in the neighbourhood of the singular point $y = y_c$, note that since $(u - c)$ and T are analytic functions of y everywhere in the finite region of the complex y-plane, they can be developed in a Taylor's series around the point $y = y_c$, as follows:

$$(u-c) = u_c'\eta + \frac{u_c''}{2!}\eta^2 + \ldots$$

$$T = T_c + T_c'\eta + \ldots \tag{19}$$

where

$$\eta = y - y_c$$

Upon substituting these series into equation (9) written in the form

$$\varphi'' - \left[\frac{T' - 2M_1^2(u-c)u'}{T - M_1^2(u-c)^2}\right]\varphi' +$$

$$+ \left[\frac{\dfrac{T' - 2M_1^2(u-c)u'}{T - M_1^2(u-c)^2}u' - u''}{u-c} - \frac{\alpha^2}{T}\left\{T - M_1^2(u-c)^2\right\}\right]\varphi = 0 \tag{20}$$

and rearranging, one obtains

$$\eta\varphi'' - \eta\left[\frac{T_c' - 2M_1^2\,u_c'^2\eta}{T_c}\right]\varphi' +$$

$$+ \left[\frac{T_c' - 2M_1^2\,u_c'^2\eta}{T_c} - \frac{u_c''}{u_c'} - \alpha^2\eta\right]\varphi = 0 \tag{21}$$

Since $\eta = 0$ is a regular singular point of equation (21), we seek a solution of equation (21) of the form (Forbenius' method)

$$\varphi_1^{(2)} = \eta^m \sum_{n=0}^{\infty} a_n\eta^n, \quad \eta > 0 \tag{22}$$

where this series is assumed *a priori* to have a nonzero radius of convergence. Substituting series (22) into equation (21), one obtains

$$\sum_{n=-1}^{\infty} a_{n+1}(m+n)(m+n+1)\,\eta^{m+n} +$$

$$-\left(\frac{T_c'}{T_c}\right)\sum_{n=0}^{\infty} a_n(m+n)\,\eta^{m+n} +$$

$$+\left(\frac{T_c'}{T_c}-\frac{u_c''}{u_c'}\right)\sum_{n=0}^{\infty} a_n\,\eta^{m+n} +$$

$$+\left(\frac{2M_1^2 u_c'^2}{T_c}\right)\sum_{n=1}^{\infty} a_{n-1}\,(m+n)\,\eta^{m+n} +$$

$$-\ \alpha^2\sum_{n=1}^{\infty} a_{n-1}\,\eta^{m+n} = 0$$

from which

$$[a_0(m-1)m]\,\eta^{m-1} + \left[a_1(m+1)m - \left(\frac{T_c'}{T_c}\right)a_0 m + \right.$$

$$\left. + \left(\frac{T_c'}{T_c}-\frac{u_c''}{u_c'}\right)a_0\right]\eta^m + \sum_{n=1}^{\infty}\left[a_{n+1}(m+n)(m+n+1) + \right.$$

$$-\left(\frac{T_c'}{T_c}\right)a_n(m+n) + \left(\frac{T_c'}{T_c}-\frac{u_c''}{u_c'}\right)a_n - \alpha^2 a_{n-1} +$$

$$\left. + \left(\frac{2M_1^2\,u_c'^2}{T_c}\right)a_{n-1}(m+n)\right]\eta^{m+n} = 0 \tag{23}$$

Equating the coefficients of each power of η to zero, one obtains, first, the indicial equation

$$a_0(m-1)m = 0 \tag{24}$$

from which, since $a_0 \neq 0$,

$$m = 1 \quad \text{or} \quad 0 \tag{25}$$

Next one obtains

$$a_1 = \frac{a_0}{2}\left[\left(\frac{T_c'}{T_c}\right) - \left(\frac{T_c'}{T_c}-\frac{u_c''}{u_c'}\right)\right]$$

or

$$a_1 = \frac{a_0}{2}\left(\frac{u_c''}{u_c'}\right) \tag{26}$$

and the recurrence relation

$$a_{n+1} = a_n \left[\frac{(m+n)(T_c'/T_c) - (T_c'/T_c - u_c''/u_c')}{(m+n)(m+n+1)}\right] + a_{n-1}\left[\frac{\alpha^2 - (2M_1^2\, u_c'^2/T_c)(m+n)}{(m+n)(m+n+1)}\right] \tag{27}$$

One thus obtains one solution:

$$\varphi_1^{(2)}(\eta) = a_0\left[\eta + \frac{1}{2}\left(\frac{u_c''}{u_c'}\right)\eta^2 + \ldots\right] \tag{28}$$

Now note that since the two roots of the indicial equation (24) differ by an integer, the other linearly-independent solution is given by

$$\varphi_2^{(2)}(\eta) = a\varphi_1^{(2)}(\eta)\ln\eta + \eta^0\left[\sum_{n=0}^{\infty} b_n\eta^n\right] \tag{29}$$

Substituting this series into equation (21) one obtains

$$\eta\left[a\left\{a_0\frac{u_c''}{u_c'} + \ldots\right\}\ln\eta + 2a\left\{a_0 + a_0\frac{u_c''}{u_c'}\eta + \ldots\right\}\frac{1}{\eta} + \right.$$
$$\left. -a\left\{a_0\eta + \frac{a_0}{2}\frac{u_c''}{u_c'}\eta^2 + \ldots\right\}\frac{1}{\eta^2} + \sum_{n=0}^{\infty} b_n\, n(n-1)\,\eta^{n-2}\right] +$$
$$-\eta\left\{\frac{T_c' - 2M_1^2\, u_c'^2\eta}{T_c}\right\}\left[a\left\{a_0 + a_0\frac{u_c''}{u_c'}\eta + \ldots\right\}\ln\eta + \right.$$
$$\left. + a_0 a\left\{\eta + \frac{1}{2}\frac{u_c''}{u_c'}\eta^2 + \ldots\right\}\frac{1}{\eta}\right] - \left\{\frac{T_c' - 2M_1^2 u_c'^2\eta}{T_c}\right\}\times$$
$$\times\sum_{n=0}^{\infty} b_n\, n\eta^n + \left\{\frac{T_c' - 2M_1^2\, u_c'^2\eta}{T_c} - \frac{u_c''}{u_c'}\right\}$$
$$\times\, a\left[a_0\eta + \frac{a_0}{2}\frac{u_c''}{u_c'}\eta^2 + \ldots\right]$$
$$+\left\{\frac{T_c' - 2M_1^2\, u_c'^2\eta}{T_c} - \frac{u_c''}{u_c'}\right\}\sum_{n=0}^{\infty} b_n\eta^n - \alpha^2\sum_{n=0}^{\infty} b_n\eta^{n+1} = 0 \tag{30}$$

Equating the coefficient of the term η^0 to zero gives

$$\left\{\frac{T_c'}{T_c} - \frac{u_c''}{u_c'}\right\} b_0 + a\, a_0 = 0$$

or

$$a = \frac{b_0}{a_0} \frac{T_c}{u_c'} \left(\frac{u'}{T}\right)'_{u=c} \tag{31}$$

Thus the two linearly-independent solutions of equation (21) are

$$\varphi_1^{(2)}(\eta) = \eta + \frac{1}{2}\left(\frac{u_c''}{u_c'}\right)\eta^2 + \ldots \tag{28}$$

$$\varphi_2^{(2)}(\eta) = \left[\frac{T_c}{u_c'}\left(\frac{u'}{T}\right)'_{u=c} \ln\eta\right] \varphi_1^{(2)}(\eta) + $$

$$+ \left[1 + \sum_{n=1}^{\infty} b_n \eta^n\right] \tag{32}$$

where the constant coefficients b_n are obtainable from equation (30).

Note that because of the singular nature of $\varphi_2^{(2)}(\eta)$, at least in the neighbourhood of y_c (where it is discontinuous and multi-valued), it cannot be a uniformly-valid asymptotic approximation to any solution of the complete viscous equations (III.22)–(III.26). Further, the main problem here is to decide the propert branch to be taken in connection with the logarithmic singularity at y_c. This reduces to the determination of the domain of the complex y-plane (excluding the neighbourhood of y_c) in which (32) is asymptotic to a solution of the complete viscous equations (III.22)–(III.26).

(3) FAR-FIELD BEHAVIOUR OF THE INVISCID DISTURBANCES

Construct a function $g(y)$

$$g(y) = \frac{\varphi(y)}{u(y) - c} \tag{33}$$

Then equation (9) becomes

$$\frac{d}{dy}\left[\left\{\frac{(u-c)^2}{T - M_1^2 (u-c)^2}\right\} g'\right] - \frac{\alpha^2 (u-c)^2}{T} g = 0 \tag{34}$$

Note that on taking the limit $\alpha \to 0$, one obtains

$$\left\{\frac{(u-c)^2}{T-M_1^2(u-c)^2}\right\} g' = \text{const.} = \text{value at a boundary}$$
$$= 0 \tag{35}$$

Therefore, g is constant between the boundary and the critical point $y = y_c$. And $g = 0$ from infinity (if $y_2 \to \infty$) down to the largest value of y_c. On the other hand, equation (34) and the boundary conditions give, as $y \to \infty$,

$$g = A\,\mathrm{e}^{-\alpha[T_1 - M_1^2(U_1-c)^2]^{1/2} y/\sqrt{T_1}} \tag{36}$$

where A is a constant, and $\alpha > 0$, however small, provided $u \to V_1$, $T \to T_1$, smoothly. Thus the order of the limits $\alpha \to 0, y \to \infty$ cannot be changed without changing the limit of the characteristic function g which, like

$$\mathrm{e}^{-\alpha[T_1 - M_1^2(V_1-c)^2]^{1/2} y/\sqrt{T_1}},$$

does not admit a power-series expansion in α uniformly for large y. In other words, corresponding to $\alpha = 0$, equation (34) exhibits a boundary layer at $y \to \infty$.

In order to study the far field, it is natural to introduce

$$Y_1 = \alpha y, \quad Y_2 = \alpha^2 y, \quad \alpha \ll 1 \tag{37}$$

and note that corresponding to the far field, one may assume for the mean flow a small departure from a uniform flow, i.e.

$$u = u(Y_2)$$
$$T = T(Y_2) \tag{38}$$

Assume, in accordance with the method of multiple scales (Van Dyke, 1975), an asymptotic expansion valid as $y \to \infty$,

$$g(y;\alpha) = g_0(Y_1, Y_2) + \alpha g_1(Y_1, Y_2) + O(\alpha^2) \tag{39}$$

so that equation (34) gives

$$\frac{\partial^2 g_0}{\partial Y_1^2} - [T_1 - M_1^2(V_1-c)^2]g_0/T_1 = 0 \tag{40}$$

$$\frac{\partial^2 g_1}{\partial Y_1^2} - [T_1 - M_1^2(V_1-c)^2]g_1/T_1$$

$$= \frac{[T_1 - M_1^2(V_1-c)^2]}{(V_1-c)^2}\left[\frac{\partial}{\partial Y_2}\left\{\frac{[u(Y_2)-c]^2}{T(Y_2) - M_1^2[u(Y_2)-c)]^2}\right\}\right]$$

$$\times \frac{\partial g_0}{\partial Y_1} + 2\,\frac{[T_1 - M_1^2(V_1-c)^2]}{(V_1-c)^2} \times$$

$$\times\left[\left\{\frac{[u(Y_2)-c]^2}{T(Y_2)-M_1^2[u(Y_2)-c]^2}\right\}\frac{\partial^2 g_0}{\partial Y_1\partial Y_2}\right] \tag{41}$$

Thus,

$$g_0=A_0(Y_2)\mathrm{e}^{-[T_1-M_1^2(V_1-c)^2]^{1/2}y_1/\sqrt{T_1}} \tag{42}$$

Substituting this into equation (41), and requiring the removal of the secular terms, one obtains

$$\left[\frac{\partial}{\partial Y_2}\left\{\frac{[u(Y_2)-c]^2}{T(Y_2)-M_1^2[u(Y_2)-c]^2}\right\}\right]A_0(Y_2)+$$

$$+\left[\frac{\{u(Y_2)-c\}^2}{T(Y_2)-M_1^2\{u(Y_2)-c\}^2}\right]\frac{\mathrm{d}A_0}{\mathrm{d}Y_2}=0$$

from which

$$A_0(Y_2)=A\left[\frac{T(Y_2)-M_1^2\{u(Y_2)-c\}^2}{\{u(Y_2)-c\}^2}\right] \tag{43}$$

where A is an arbitrary constant.

Thus,

$$g(y;\alpha)=A\left[\frac{T(\alpha^2y)-M_1^2\{u(\alpha^2y)-c\}^2}{\{u(\alpha^2y)-c\}^2}\right]\times$$

$$\times\,\mathrm{e}^{-\alpha[T_1-M_1^2(V_1-c)^2]^{1/2}y/\sqrt{T_1}}+O(\alpha) \tag{44}$$

5

An Initial-value-problem Approach

INTRODUCTION

Here we pose an initial-value problem for the linearised-inviscid-disturbance equation to obtain a prescription for the proper path of integration across the critical point $y = y_c$, associated with the inviscid solutions.

(1) THE INITIAL-VALUE PROBLEM

In the normal-mode approach, each disturbance is resolved into dynamically-independent components, and each mode of perturbation is assumed to be of the form $e^{i\alpha(x-ct)}$ times an amplitude function of y. One then searches for those values of c for which the linearised equations of the flow with appropriate boundary conditions possess nontrivial solutions. If there are admissible values of c with $\mathrm{Im}(c) > 0$, the perturbed basic flow is said to be unstable. Otherwise, it is stable.

One may seek to justify the normal-mode procedure by following the subsequent time evolution of a small perturbation introduced, say, at $t = 0$, through a normal-mode expansion (assuming that the normal modes are complete). If any of the normal modes have $\mathrm{Im}(c) > 0$, the perturbation would grow, and one may then claim instability in the physical sense. But this also indicates that the problem of hydrodynamic stability is really an initial-value problem, and should be, strictly speaking, posed as such – as recognised by Case (1960), and Dikii (1961). In particular, one really wants to know whether an initial perturbation grows in time. In most problems the initial-value-problem approach gives results that complement those due to the normal-mode approach. It turns out, for instance, that in an initial-value-problem analysis, one cannot miss the modes associated with a continuous spectrum of characteristic values, unlike that in a normal-mode analysis. But, a notable merit of the initial-value-problem approach is its ability to resolve some of the difficulties that arise (as we saw in Chapter III) in the course of a normal-mode analysis, without having to go outside the inviscid model for a gas.

Such considerations are not peculiar to fluid dynamics. A classical example in other fields is that of longitudinal plasma oscillations in plasma physics. The ambiguities that arise in a straightforward Fourier-transform approach are resolvable readily when one poses the issue as an initial-value problem (Landau, 1946).

Just to keep the algebra simple, in the following we restrict ourselves to very small Mach numbers, $M_1 \ll 1$ (Shivamoggi, 1982). The extension of the analysis to arbitrary Mach numbers is straightforward. Equation (IV.9) then becomes

$$(\varphi'' - \alpha^2 \varphi) - \left(\frac{T'}{T}\right) \varphi' - \frac{T}{u - c} \left(\frac{u'}{T}\right)' \varphi = 0 \tag{1}$$

with the boundary conditions

$$y = y_1, y_2 : \varphi = 0 \tag{2}$$

The initial-value problem corresponding to (1) and (2) is

$$\left(u - \frac{\mathrm{i}}{\alpha} \frac{\partial}{\partial t}\right) \left(\frac{\partial^2}{\partial y^2} - \alpha^2\right) \varphi - \frac{T'}{T} \left(u - \frac{\mathrm{i}}{\alpha} \frac{\partial}{\partial t}\right) \varphi' +$$

$$- T \left(\frac{u'}{T}\right)' \varphi - 0 \tag{3}$$

$$y = y_1, y_2 : \varphi = 0 \tag{4}$$

$$t = 0 : \varphi = \varphi_0 \tag{5}$$

Upon Laplace transforming according to

$$\bar{\varphi}(y) = \int_0^\infty \mathrm{e}^{-st} \varphi(y, t)\, \mathrm{d}t$$

(3)–(5) give

$$\left(u - \frac{\mathrm{i}s}{\alpha}\right) \left(\frac{\mathrm{d}^2}{\mathrm{d}y^2} - \alpha^2\right) \bar{\varphi} - \frac{T'}{T} \left(u - \frac{\mathrm{i}s}{\alpha}\right) \frac{\mathrm{d}\bar{\varphi}}{\mathrm{d}y} +$$

$$- T \left(\frac{u'}{T}\right)' \bar{\varphi} = g(y) \tag{6}$$

where

$$g(y) \equiv -\frac{i}{\alpha}(\varphi_0'' - \alpha^2 \varphi_0) + \frac{T'}{T}\frac{i}{\alpha}\varphi_0'$$

One may construct an explicit solution to (6), using the Green-function method:

$$\bar{\varphi}(y) = \int_{y_1}^{y} \frac{\bar{\psi}_1(y)\,\bar{\psi}_2(\xi)}{\left[u(\xi) - \frac{is}{\alpha}\right]\bar{W}(\bar{\psi}_1, \bar{\psi}_2)}\, g(\xi)\,d\xi + \int_{y}^{y_2} \frac{\bar{\psi}_1(\xi)\,\bar{\psi}_2(y)}{\left[u(\xi) - \frac{is}{\alpha}\right]\bar{W}(\bar{\psi}_1, \bar{\psi}_2)}\, g(\xi)\,d\xi \tag{7}$$

where

$$\left[\left(u - \frac{is}{\alpha}\right)\left(\frac{d^2}{dy^2} - \alpha^2\right) - \frac{T'}{T}\left(u - \frac{is}{\alpha}\right)\frac{d}{dy} + {} - T\left(\frac{u'}{T}\right)'\right]\bar{\psi}_{1,2} = 0$$

$$\bar{\psi}_1(y_1) = 0, \quad \bar{\psi}_2(y_2) = 0$$

$$\bar{W}(\bar{\psi}_1, \bar{\psi}_2) \equiv \bar{\psi}_1(y)\,\bar{\psi}_2'(y) - \bar{\psi}_1'(y)\,\bar{\psi}_2(y)$$

One may construct $\bar{\psi}_1$, $\bar{\psi}_2$ from the two linearly-independent solutions $\bar{\varphi}_1, \bar{\varphi}_2$ as follows:

$$\bar{\psi}_1 = \bar{\varphi}_1(y)\bar{\varphi}_2(y_1) - \bar{\varphi}_1(y_1)\bar{\varphi}_2(y)$$

$$\bar{\psi}_2 = \bar{\varphi}_1(y)\bar{\varphi}_2(y_2) - \bar{\varphi}_1(y_2)\bar{\varphi}_2(y)$$

Thus,

$$\bar{W}(\bar{\psi}_1, \bar{\psi}_2) = -\bar{W}(\bar{\varphi}_1, \bar{\varphi}_2)\cdot\Delta$$

where

$$\Delta = \begin{vmatrix} \bar{\varphi}_1(y_1) & \bar{\varphi}_2(y_1) \\ \bar{\varphi}_1(y_2) & \bar{\varphi}_2(y_2) \end{vmatrix}$$

The solution to (3)–(5) is then given on inversion,

$$\varphi(y, t) = \frac{1}{2\pi i}\int_{-i\infty+\delta}^{i\infty+\delta} \bar{\varphi}(y)\, e^{st}\,ds \tag{8}$$

The path of integration in (8) corresponds to the Bromwich contour, which lies to the right of all the singularities of $\bar{\varphi}(y)$. In the absence of singularities other than the one at $u = is/\alpha$ on the imaginary axis, one may push the path of integration along the imaginary axis ($\delta \to 0$), and continue around the pole at $u = is/\alpha$ from the right, if $u_c' > 0$, and vice versa.

In the normal-mode formulation, $s \to -i\alpha c$, this would imply that the path of integration is continued around the critical point $y = y_c$ (when c is real) from below the real axis in the complex y-plane when $u_c' > 0$, and vice versa, (see Fig. 1).

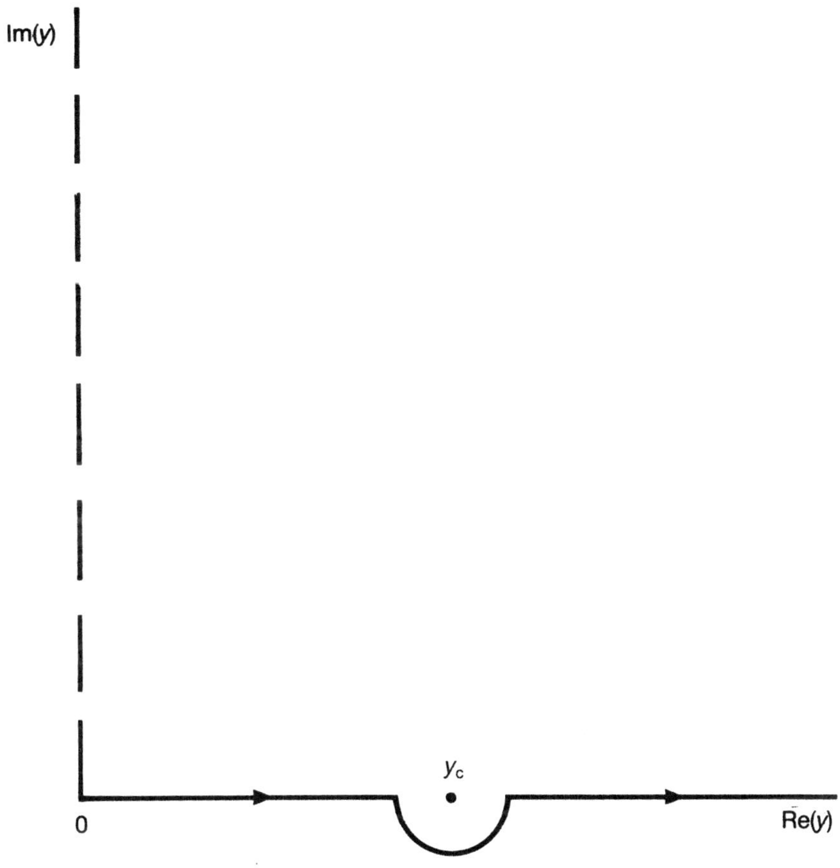

Fig. 5.1

Other poles of $\bar{\varphi}(y)$ arise at

$$\bar{W}(\bar{\psi}_1, \bar{\psi}_2) = 0 \quad \text{or } \Delta = 0$$

$$\text{for } \mathrm{Re}(s) > 0$$

which is simply the characteristic-value relation for the discrete spectrum of normal modes with $\mathrm{Im}(c) > 0$.

Thus, one obtains

$$\varphi(y,t) = \frac{1}{2\pi i}\int_{-i\infty+\delta}^{i\infty+\delta} \bar{\varphi}(y)\, e^{st}\, \varphi s +$$

$$+ \sum_{\substack{\text{discrete} \\ \text{spectrum}}} (\text{exponentials growing like } e^{\alpha[\text{Im}(c)]t}) \qquad (9)$$

Consider the asymptotic behaviour of the first term on the right-hand side in (9). The integral picks up contributions from the poles at $u = is/\alpha = c$. Recall from Chater IV that φ_1, φ_2 in the neighbourhood of the singular point $y = y_c$ are given by

$$\varphi_1(y) = (y - y_c) + \frac{1}{2}\left(\frac{u_c''}{u_c'}\right)(y - y_c)^2 + \ldots$$

$$\varphi_2(y) = \left[\frac{T_c}{u_c'}\left(\frac{u'}{T}\right)'_{u=c} \ln(y - y_c)\right]\varphi_1(y) +$$

$$+ \left[1 + \sum_{n=1}^{\infty} b_n (y - y_c)^n\right] \qquad (10)$$

The contributions to the integral, thus, come from a simple pole at $u = c$, and the logarithmic term in φ_2. Upon inversion, this simple pole produces a simple exponential $e^{i\alpha ct}$ (c real), while the logarithm produces a term behaving like $1/t$. Therefore, for large t, the dominant contribution to the solution will come from the discrete spectrum, if there is one, from the term growing like $e^{\alpha[\text{Im}(c)]t}$ for the largest value of $\alpha[\text{Im}(c)]$, with $\text{Im}(c) > 0$. This means that the continuous spectrum does not correspond to instability, and the discrete spectrum alone is associated with instability. Thus, in seeking a criterion for instability, one may use the method of normal modes, and ignore the continuous spectrum.

6

Inviscid-stability Theory

INTRODUCTION

We proceed to discuss here a general theory of linearised inviscid disturbance in parallel compressible flows (Shivamoggi, 1977a). First, local and global necessary conditions for the existence of nonneutral and neutral modes of 'subsonic' disturbances are established. A semi-circle theorem, viz. that the complex wave velocity for a self-excited 'subsonic' mode lies inside the semi-circle in the upper-half c-plane which has the range of u for its diameter, is established. We shall discuss various kinds of possible neutral 'subsonic' modes. Toward establishing the sufficient conditions for the existence of self-excited 'subsonic' modes, we first exhibit the variational character of the inviscid problem for monotonic symmetrical type of mean flow, and then use the Sturm–Liouville theorem to ensure the existence of a regular neutral 'subsonic' mode, and then construct unstable solutions that are contigous with the neutral modes. We shall further establish that the number of distinct unstable 'subsonic' modes cannot exceed the number of netural 'subsonic' modes.

(1) INVISCID PROBLEM

Let us recapitulate the inviscid problem given by

$$\frac{\mathrm{d}}{\mathrm{d}y}\left[\frac{\varphi'}{T-M_1^2(u-c)^2}\right]-\left[\frac{\left\{\dfrac{u'}{T-M_1^2(u-c)^2}\right\}'}{u-c}+\frac{\alpha^2}{T}\right]\varphi=0 \qquad (1)$$

$$y = y_1, y_2 : \varphi = 0 \tag{2}$$

An examination of equation (1) shows that it is natural to classify the inviscid disturbances as subsonic, sonic and supersonic, according as the phase velocity in the direction of the mean flow and relative to an observer moving with the velocity of the mean flow is – in magnitude – below, equal to, or above the local speed of sound, i.e.

subsonic disturbances: $M_1^2(u-c)^2 < T$
supersonic disturbances: $M_1^2(u-c)^2 > T$
sonic disturbances: $M_1^2(u-c)^2 = T$

(2) DISCONTINUITIES IN THE MEAN FLOW

If u or u' is discontinuous at y_0, say, then the pressure must be continuous at the material–fluid interface with mean position at $y = y_0$. Therefore, noting that

$$(u-c)\left\{\frac{\varphi'}{T-M_1^2(u-c)^2}\right\}' - \left\{\frac{u'}{T-M_1^2(u-c)^2}\right\}'\varphi$$
$$= \left\{(u-c)\left[\frac{\varphi'}{T-M_1^2(u-c)^2}\right] - \left[\frac{u'}{T-M_1^2(u-c)^2}\right]\varphi\right\}'$$

one has,

$$\left[(u-c)\left\{\frac{\varphi'}{T-M_1^2(u-c)^2}\right\} - \left\{\frac{u'}{T-M_1^2(u-c)^2}\right\}\varphi\right] = 0 \tag{3}$$

at $y = y_0$, where the square brackets denote the jump of their contents at such a possible discontinuity. Also the normal velocity of the fluid must be continuous at the material interface. Let this interface be represented by

$$y = y_0 + \xi(x, y) \tag{4}$$

where

$$\xi(x, t) = F_0\, \mathrm{e}^{\mathrm{i}\alpha(x-ct)}$$

Then

$$v' = \frac{\mathrm{D}\xi}{\mathrm{D}t} = \frac{\partial\xi}{\partial t} + u\frac{\partial\xi}{\partial x} = \mathrm{i}\alpha(u-c)\xi \tag{5}$$

Noting $v' = \alpha\varphi$ one obtains

$$\left[\frac{\varphi}{u-c}\right] = 0 \tag{6}$$

at $y = y_0$.

Explicit solutions of the characteristic-value problem of (1), (2) are difficult to find in practice for smoothly-varying functions $u(y)$, $T(y)$. However, when $u(y)$ is a piecewise-linear function, the situation is tractable, and one can then join up such solutions by using conditions (3) and (6) – a method first used by Kelvin (1887) for the case of an incompressible fluid. As an illustration, consider the Kelvin–Helmholtz instability problem of two uniform fluids in relative horizontal motion separated by a horizontal boundary.

Example. Let two uniform fluids of densities ρ_1, ρ_2 (temperatures T_1 and T_2) and velocities V_1, V_2 be separated by a horizontal boundary at $y = 0$. One has

$$\varphi'' - \frac{\alpha^2}{T_1}\{T_1 - M_1^2(V_1 - c)^2\}\,\varphi = 0, \quad y > 0$$

$$\varphi'' - \frac{\alpha^2}{T_2}\{T_2 - M_1^2(V_2 - c)^2\}\,\varphi = 0, \quad y < 0 \tag{7}$$

From which, noting (2) and (6), one has

$$\varphi_1 = A(V_1 - c)\,e^{-s_1 y}, \quad y > 0$$

$$\varphi_2 = A(V_2 - c)\,e^{+s_2 y}, \quad y < 0 \tag{8}$$

as the solutions appropriate for the two regions, where

$$s_1 = \frac{\alpha}{\sqrt{T_1}}\{T_1 - M_1^2(V_1 - c)^2\}^{1/2}$$

$$s_2 = \frac{\alpha}{\sqrt{T_2}}\{T_2 - M_1^2(V_2 - c)^2\}^{1/2}$$

and A is an arbitrary constant.

Using (8) in (3), one finds the characteristic-value equation

$$\frac{(V_1 - c)^2}{\sqrt{T_1}\{T_1 - M_1^2(V_1 - c)^2\}^{1/2}} + \frac{(V_2 - c)^2}{\sqrt{T_2}\{T_2 - M_1^2(V_2 - c)^2\}^{1/2}} = 0 \tag{9}$$

Putting

$$\beta_1 = \frac{\sqrt{T_1}\{T_1 - M_1^2(V_1 - c)^2\}^{1/2}}{\sqrt{T_1}\{T_1 - M_1^2(V_1 - c)^2\}^{1/2} + \sqrt{T_2}\{T_2 - M_1^2(V_2 - c)^2\}^{1/2}}$$

$$\beta_2 = \frac{\sqrt{T_2}\{T_2 - M_1^2(V_2 - c)^2\}^{1/2}}{\sqrt{T_1}\{T_1 - M_1^2(V_1 - c)^2\}^{1/2} + \sqrt{T_2}\{T_2 - M_1^2(V_2 - c)^2\}^{1/2}} \tag{10}$$

Equation (9) then becomes

$$\beta_2\left(V_1 + \frac{n}{\alpha}\right)^2 + \beta_1\left(V_2 + \frac{n}{\alpha}\right)^2 = 0 \tag{11}$$

where

$$c = -\frac{n}{\alpha}$$

Expanding equation (11),

$$n^2 - 2\alpha(\beta_1 V_2 + \beta_2 V_1) + \alpha^2(\beta_1 V_2^2 + \beta_2 V_1^2) = 0 \tag{12}$$

The roots of this equation are given by

$$n = \tfrac{1}{2}(\beta_1 V_2 + \beta_2 V_1) \pm \mathrm{i}\alpha \frac{\sqrt{\beta_1\beta_2}}{2} \,|V_1 - V_2| \tag{13}$$

Thus in the absence of a stabilising external body-force distribution such as gravity, such an interface appears always to be unstable corresponding to small Mach numbers M_1.

(3) ODD AND EVEN SOLUTIONS

Next the solutions for the disturbances exhibit certain symmetries when the basic flow is symmetric, i.e. say $y_1 = -y_2$, and $u(y)$ is an even function. In that case, if $\varphi(y)$ is a characteristic function with a characteristic value c for a given α, it follows that the even part:

$$\varphi_e = \tfrac{1}{2}[\varphi(y) + \varphi(-y)] \tag{14}$$

and the odd part:

$$\varphi_o = \tfrac{1}{2}[\varphi(y) - \varphi(-y)] \tag{15}$$

of φ are also characteristic functions for the same c, α. It can be shown further that either φ_o of φ_e is identically zero. In order to see this, multiply equation (1) for φ_e by φ_o and subtract φ_e times the corresponding equation for φ_o. Then one obtains

$$\varphi_e\left\{\frac{\varphi_o'}{T - M_1^2(u-c)^2}\right\}' - \varphi_o\left\{\frac{\varphi_e'}{T - M_1^2(u-c)^2}\right\}' = 0 \tag{16}$$

where $u \neq c$. Therefore,

$$\varphi_e\left\{\frac{\varphi_o'}{T - M_1^2(u-c)^2}\right\} - \varphi_o\left\{\frac{\varphi_e'}{T - M_1^2(u-c)^2}\right\} = \text{const.} = 0 \tag{17}$$

Thus, in general, φ_o and φ_e are linearly dependent, which is possible only if one of them is identically zero. Thus it follows that a characteristic function is either odd or even, except when c is real.

In fact, both even and odd characteristic functions can be found for the same basic symmetric flow, but they have different characteristic values c for each α. The oddness (or evenness) of φ allows one to reduce the effective field of flow to the half domain, $0 \leqslant y \leqslant y_2$, applying the symmetry condition $\varphi = 0$ (or $\varphi' = 0$) at $y = 0$ and the original boundary condition $\varphi = 0$ at $y = y_2$.

Next suppose that the basic flow profile is antisymmetric, with $y_1 = -y_2$, odd $u(y)$. Then for each characteristic function φ with a characteristic value c, there is a characteristic function $\varphi^*(-y)$ with a characteristic value $c^* = -c_r + ic_i$ for the same α. When the characteristic solution is unique, the Hermitian symmetry implies that $c_r = 0$ and $\varphi^*(-y) = \varphi(y)$. Otherwise, there may be a pair of characteristic solutions with phase velocities $\pm c_r(\alpha)$ and the same $c_i(\alpha)$, one function being the Hermitian conjugate of the other.

(4) SELF-EXCITED AND DAMPED DISTURBANCES

It would appear that for the inviscid problem (1), (2) it is meaningless to distinguish between self-excited and damped disturbances, because for each characteristic function with characteristic value c, for a given α, there is another complex conjugate characteristic function φ^* with characteristic value $c^* = c_r - ic_i$ for the same α. (This is evident by taking the complex conjugate of (1), (2).) Thus to each amplified wave there would be a damped wave, and vice versa, which merely signifies the time symmetry of the problem, comprising periodic motion of inviscid fluid with steady boundaries. However, this is not correct since the inviscid equations have meaning only as the limiting case of the complete viscous equations, and the latter are not invariant under complex conjugation, unlike the inviscid problem.

(5) LOCAL AND GLOBAL NECESSARY CONDITIONS FOR THE EXISTENCE OF NONNEUTRAL AND NEUTRAL SUBSONIC MODES

THEOREM: *If the flow possesses a nonneutral mode of subsonic disturbance with a finite wavelength, the velocity profile and the temperature profile satisfy at some point $y = y_s$, where $y_1 \leqslant y \leqslant y_2$, a local necessary condition,*

$$\left\{\frac{u'}{T - M_1^2 (u-c)^2}\right\}'_{y=y_s} = 0$$

where we restrict ourselves to $|c_i| \ll 1$, and $M_1 \ll 1$. Furthermore in the case of a neutral mode, one must have $u(y_s) = c$, $(c = c_r)$.

Proof: Multiply equation (1) by φ^*, the complex conjugate of φ, and integrate from y_1 to y_2; then using (2), one obtains

$$-\int_{y_1}^{y_2}\left[\left\{\frac{|\varphi'|^2}{T-M_1^2(u-c)^2}\right\}+\frac{\alpha^2}{T}|\varphi|^2\right]\mathrm{d}y$$

$$=\int_{y_1}^{y_2}\frac{\dfrac{\mathrm{d}}{\mathrm{d}y}\left\{\dfrac{u'}{T-M_1^2(u-c)^2}\right\}}{u-c}|\varphi|^2\,\mathrm{d}y \tag{18}$$

The imaginary part of equation (18), for the case $|c_i| \ll 1, M_1 \ll 1$, gives

$$c_i\int_{y_1}^{y_2}\frac{\dfrac{\mathrm{d}}{\mathrm{d}y}\left\{\dfrac{u'}{T-M_1^2(u-c)^2}\right\}}{|u-c|^2}|\varphi|^2\,\mathrm{d}y=0 \tag{19}$$

If $c_i \neq 0$, it follows that

$$\frac{\mathrm{d}}{\mathrm{d}y}\left\{\frac{u'}{T-M_1^2(u-c)^2}\right\}$$

must change sign at one or more points in the flow field. Assuming that

$$\frac{\mathrm{d}}{\mathrm{d}y}\left\{\frac{u'}{T-M_1^2(u-c)^2}\right\}$$

is continuous for $y_1 \leqslant y \leqslant y_2$, there must be at least one point in the interval (y_1, y_2) where, for subsonic disturbances, one has

$$\frac{\mathrm{d}}{\mathrm{d}y}\left\{\frac{u'}{T-M_1^2(u-c)^2}\right\}_{y=y_s}=0 \tag{20}$$

Next in order to show that in the case of a neutral mode one must have $u(y_s) = c$ $(c = c_r)$, multiply equation (1) written in the form

$$\varphi''-\left[\frac{T'-2M_1^2(u-c)u'}{T-M_1^2(u-c)^2}\right]\varphi'+$$

$$+\left[\frac{\dfrac{T'-2M_1^2(u-c)u'}{T-M_1^2(u-c)^2}u'-u''}{u-c}-\frac{\alpha^2}{T}\{T-M_1^2(u-c)^2\}\right]\varphi=0 \tag{21}$$

by φ^*, and subtract from the resulting equation its complex conjugate, and then there follows for $|c_i| \ll 1, M_1 \ll 1$,

$$\frac{\mathrm{d}}{\mathrm{d}y}(\varphi'\varphi^* - \varphi^{*\prime}\varphi) - \left[\frac{T' - 2M_1^2(u-c)u'}{T - M_1^2(u-c)^2}\right]\varphi'\varphi^* +$$

$$+\left[\frac{T' - 2M_1^2(u-c)u'}{T - M_1^2(u-c)^2}\right]\varphi^{*\prime}\varphi +$$

$$+(c - c^*)\left[\frac{\dfrac{T' - 2M_1^2(u-c)u'}{T - M_1^2(u-c)^2}u' - u''}{|u-c|^2}\right]|\varphi|^2 = 0 \tag{22}$$

Introducing the 'Wronskian'

$$W \equiv \frac{\mathrm{i}}{2}\,\frac{(\varphi\varphi^{*\prime} - \varphi'\varphi^*)}{T - M_1^2(u-c)^2} \tag{23}$$

equation (22) gives

$$\frac{\mathrm{d}W}{\mathrm{d}y} - c_{\mathrm{i}}\left[\frac{\left\{\dfrac{u'}{T - M_1^2(u-c)^2}\right\}'}{|u-c|^2}\right]|\varphi|^2 = 0 \tag{24}$$

If we let $c_{\mathrm{i}} \to 0$ in equation (24), we obtain results corresponding to neutral disturbances. Thus

$$W = \text{const.} \tag{25}$$

except possibly at the critical point y_c (where $u = c$, $c = c_{\mathrm{r}}$) which may now lie in the real interval (y_1, y_2). If it does (we shall soon show that it must), W has a discontinuity at the point. Integrate equation (24) from y_c^- to y_c^+, then

$$W(y_c^+) - W(y_c^-) = \int_{u_c^-}^{u_c^+} \frac{\left\{\dfrac{u'}{T - M_1^2(u-c)^2}\right\}'|\varphi|^2\, c_{\mathrm{i}}\,\mathrm{d}u}{u'\{(u - c_{\mathrm{r}})^2 + c_{\mathrm{i}}^2\}} \tag{26}$$

Note that this is of the form

$$I(x, 0^+) = \mathop{\mathrm{Lt}}_{\substack{y \to 0^+ \\ \epsilon \to 0}} \int_{x-\epsilon}^{x+\epsilon} \frac{h(\xi)}{y\left[1 + \dfrac{(x-\xi)^2}{y^2}\right]}\,\mathrm{d}\xi$$

Let

$$\mu = \frac{x - \xi}{y}$$

so that

$$I(x,0^+) = h(x) \underset{\substack{y\to 0^+\\ \epsilon\to 0}}{\mathrm{Lt}} \int_{-\epsilon/y}^{\epsilon/y} \frac{\mathrm{d}\mu}{1+\mu^2}$$

or

$$I(x,0^+) = 2h(x) \underset{\substack{y\to 0^+\\ \epsilon\to 0}}{\mathrm{Lt}} \left[\tan^{-1}\frac{\epsilon}{y}\right]$$

Taking the limits $y \to 0^+$, $\epsilon \to 0$ in such a way that $\epsilon/y \to \infty$, one finds

$$I(x,0^+) = \pi h(x)$$

Thus as $c_i \to 0$, equation (26) gives

$$W(y_c^+) - W(y_c^-) \equiv [W]_c = \pi \frac{(u'/T)'_{u=c}}{u'_c} |\varphi_c|^2 \tag{27}$$

provided in this limit, $c_i > 0$ (recall from Chapter V that only this view led to a proper resolution of some essential difficulties), and $u_c^+ > u_c^-$. Equation (27) shows that if a neutral subsonic mode is to exist, there are two possibilities:

(1) either all the jumps $[W]_c = 0$, i.e. $(u'/T)' = 0$ at all points where $u = c$ ($c = c_r$);

(2) or several of the jumps $[W]_c$ cancel out, then $(u'/T)'$ changes sign at the various points where $u = c$ ($c = c_r$).

If the mean-flow profiles are monotonic, then the first possibility prevails. In this case, equation (1) has regular solutions which are physically possible in the absence of viscous effects.

A global necessary condition

We next deduce a global necessary condition for the existence of a nonneutral or neutral subsonic mode.

The real part of equation (18) gives

$$\int_{y_1}^{y} (u - c_r) \frac{\dfrac{\mathrm{d}}{\mathrm{d}y}\left\{\dfrac{u'}{T - M_1^2(u-c)^2}\right\}}{|u-c|^2} |\varphi|^2 \,\mathrm{d}y$$

$$= - \int_{y_1}^{y_2} \left[\left\{\frac{|\varphi'|^2}{T - M_1^2(u-c)^2}\right\} + \frac{\alpha^2}{T}|\varphi|^2\right] \mathrm{d}y \tag{28}$$

Adding

$$(c_r - u_s) \int_{y_1}^{y_2} \frac{\dfrac{d}{dy}\left\{\dfrac{u'}{T - M_1^2(u-c)^2}\right\}}{|u-c|^2} |\varphi|^2 \, dy = 0$$

to the left-hand side of equation (28), one obtains

$$\int_{y_1}^{y_2} \frac{\left[\dfrac{d}{dy}\left\{\dfrac{u'}{T - M_1^2(u-c)^2}\right\}\right](u-u_s)}{|u-c|^2} |\varphi|^2 \, dy$$

$$= - \int_{y_1}^{y_2} \left[\left\{\frac{|\varphi'|^2}{T - M_1^2(u-c)^2}\right\} + \frac{\alpha^2}{T}|\varphi|^2\right] dy \qquad (29)$$

From which one obtains a global necessary condition for the existence of a non-neutral or netural subsonic mode:

$$\left[\frac{d}{dy}\left\{\frac{u'}{T - M_1^2(u-c)^2}\right\}\right](u - c_s) < 0 \qquad (30)$$

where, for $y_1 \leqslant y_s \leqslant y_2$,

$$\frac{d}{dy}\left\{\frac{u'}{T - M_1^2(u-c)^2}\right\}_{y=y_s} = 0 \qquad (20)$$

and

$$u_s = u(y_s)$$

In particular, if $u(y)$, $T(y)$ are monotonic functions and

$$\frac{d}{dy}\left\{\frac{u'}{T - M_1^2(u-c)^2}\right\}$$

vanishes only once in the flow field, $y_1 \leqslant y \leqslant y_2$, a combined local and global necessary condition for instability is that

$$\left[\frac{d}{dy}\left\{\frac{u'}{T - M_1^2(u-c)^2}\right\}\right](u - c_s) \leqslant 0 \qquad (31)$$

throughout the flow field, $y_1 \leqslant y \leqslant y_2$, with equality valid only where $y = y_s$ (this result being valid only for subsonic modes). The counterpart of this result for the case of an incompressible fluid was established by Fjortoft (1950).

(6) A SEMI-CIRCLE THEOREM

We next deduce a semi-circle theorem for nonneutral disturbances. First construct a function $g(y)$ which is supposed to be regular in the interval (y_1, y_2):

$$g(y) = \frac{\varphi(y)}{u(y) - c} \tag{32}$$

Then equation (1) becomes

$$\frac{\mathrm{d}}{\mathrm{d}y}\left[\left\{\frac{(u-c)^2}{T - M_1^2(u-c)^2}\right\} g'\right] - \frac{\alpha^2(u-c)^2}{T} g = 0 \tag{33}$$

Multiply equation (32) by g^*, g^* being the complex conjugate of g, and integrate between (y_1, y_2); then

$$\int_{y_1}^{y_2} \frac{(u-c)^2}{T - M_1^2(u-c)^2} \left\{ |g'|^2 + \frac{\alpha^2}{T} [T - M_1^2(u-c)^2] |g|^2 \right\} \mathrm{d}y = 0 \tag{34}$$

where we have transformed the first part by partial integration.

THEOREM: *If $c_i \neq 0$, c_r must lie within the range of u.*

Proof: The real and imaginary parts of equation (34) for the case $M_1 \ll 1$ give

$$\int_{y_1}^{y_2} \frac{(u-c_r)^2 - c_i^2}{T - M_1^2(u-c)^2} \left\{ |g'|^2 + \frac{\alpha^2}{T} [T - M_1^2(u-c)^2] |g|^2 \right\} \mathrm{d}y = 0 \tag{35}$$

and

$$c_i \int_{y_1}^{y_2} \frac{(u-c_r)}{T - M_1^2(u-c)^2} \left\{ |g'|^2 + \frac{\alpha^2}{T} [T - M_1^2(u-c)^2] |g|^2 \right\} \mathrm{d}y = 0 \tag{36}$$

If we restrict ourselves to subsonic disturbances, equation (36) shows that c_r must lie within the range of u if $c_i \neq 0$ – a result whose counterpart for the case of an incompressible fluid was established by Synge (1933).

THEOREM: *The complex wave velocity for a self-excited subsonic mode lies inside the semi-circle in the upper half of the c-plane which has the range of u for its diameter.*

Proof: From equation (36), for $c_i \neq 0$,

$$\int_{y_1}^{y_2} \frac{u}{T - M_1^2(u-c)^2} \left\{ |g'|^2 + \frac{\alpha^2}{T} [T - M_1^2(u-c)^2] |g|^2 \right\} dy$$
$$= \int_{y_1}^{y_2} \frac{c_r}{T - M_1^2(u-c)^2} \left\{ |g'|^2 + \frac{\alpha^2}{T} [T - M_1^2(u-c)^2] |g|^2 \right\} dy \tag{37}$$

Using equation (37), equation (35) becomes

$$\int_{y_1}^{y_2} \frac{u^2}{T - M_1^2(u-c)^2} \left\{ |g'|^2 + \frac{\alpha^2}{T} [T - M_1^2(u-c)^2] |g|^2 \right\} dy$$
$$= \int_{y_1}^{y_2} \frac{c_r^2 + c_i^2}{T - M_1^2(u-c)^2} \left\{ |g'|^2 + \frac{\alpha^2}{T} [T - M_1^2(u-c)^2] |g|^2 \right\} dy \tag{38}$$

Note the inequality

$$0 \geqslant \int_{y_1}^{y_2} (u - u_{\min})(u - u_{\max}) \times$$
$$\times \frac{\left\{ |g'|^2 + \frac{\alpha^2}{T} [T - M_1^2(u-c)^2] |g|^2 \right\}}{\{T - M_1^2(u-c)^2\}} dy \tag{39}$$

Using equations (37) and (38), this becomes

$$\int_{y_1}^{y_2} \frac{\left\{ |g'|^2 + \frac{\alpha^2}{T} [T - M_1^2(u-c)^2] |g|^2 \right\}}{\{T - M_1^2(u-c)^2\}} \times$$
$$\times [(c_r^2 + c_i^2) - (u_{\min} + u_{\max}) c_r + u_{\min} u_{\max}] dy \leqslant 0 \tag{40}$$

Noting that

$$\frac{\left\{ |g'|^2 + \frac{\alpha^2}{T} [T - M_1^2(u-c)^2] |g|^2 \right\}}{\{T - M_1^2(u-c)^2\}} \geqslant 0$$

one then finds

$$c_r^2 + c_i^2 - (u_{\min} + u_{\max}) c_r + u_{\min} u_{\max} \leqslant 0$$

or

$$[c_r - \tfrac{1}{2}(u_{\min} + u_{\max})]^2 + c_i^2 \leqslant \tfrac{1}{2} [\tfrac{1}{2}(u_{\max} - u_{\min})]^2 \tag{41}$$

which shows that the complex wave velocity for a self-excited subsonic mode must lie inside the semi-circle in the upper half of the complex c-plane which has the range of u for its diameter along the real axis – a result whose counterpart for the case of an incompressible fluid was established by Howard (1961). Compressibility effects do not seem to influence the result, at least for subsonic modes.

(7) SPECIAL NEUTRAL MODES

THEOREM: *If c is real (i.e. $c = c_r$) and falls within the range of u, and $(u'/T)'_{u=c} \neq 0$, then a singular neutral subsonic mode is possible, i.e.*

$$\varphi(y_c) = \varphi_c \neq 0$$

Proof: If $(u'/T)' \neq 0$ at all points where $u = c$, the neutral subsonic mode is given by

$$\varphi(y) = A\varphi_1(y) + B\varphi_2(y)$$

where A and B are arbitrary constants and

$$\varphi_1(y) = (y - y_c) + \frac{1}{2}\left(\frac{u_c''}{u_c'}\right)(y - y_c)^2 + \ldots$$

$$\varphi_2(y) = \left[\frac{T_c}{u_c'}\left(\frac{u'}{T}\right)'_{u=c} \ln(y - y_c)\right]\varphi_1(y) +$$

$$+ \left[1 + \sum_{n=1}^{\infty} b_n (y - y_c)^n\right]$$

and the latter is not regular, Assume first that

$$\varphi_c = 0$$

Then, $B = 0$, so that

$$\varphi(y) = A\varphi_1(y)$$

and φ' exists, even at y_c. Integration of equation (1) between y and y_c gives

$$(u - c)\left\{\frac{\varphi'}{T - M_1^2(u - c)^2}\right\} - \left\{\frac{u'}{T - M_1^2(u - c)^2}\right\}\varphi$$

$$= \alpha^2 \int_{y_c}^{y} \frac{(u - c)}{T}\varphi \, dy \qquad (42)$$

where both c and φ are real. Rearranging equation (42), one obtains

$$-\frac{\varphi^2}{T-M_1^2(u-c)^2}\left(\frac{u-c}{\varphi}\right)' = \alpha^2 \int_{y_c}^{y} \frac{(u-c)}{T}\,\varphi\,\mathrm{d}y \qquad (43)$$

Since a change of sign for φ is always allowable, assume, for $y > y_c$,

$$\left(\frac{u-c}{T}\right)\varphi > 0 \qquad (44)$$

Then, equation (43) gives for $y > y_c$

$$-\left(\frac{u-c}{\varphi}\right)' > 0 \qquad (45)$$

Repeating this argument one finds that

$$\left(\frac{u-c}{\varphi}\right) > 0 \qquad (46)$$

i.e. $(u - c)$ and φ are always of the same sign, and that $(u-c)/\varphi$ decreases as y increases, so that φ cannot vanish at $y = y_2$, as it should: therefore the contradiction, and so

$$\varphi_c \neq 0$$

THEOREM: *A special regular neurtral subsonic mode, with wavenumber zero, is possible, if u has the same value at the boundaries* y_1, y_2.

Proof: Inspection of equation (1) shows that

$$\alpha = 0 : \varphi = u - c, \qquad c = c_r \qquad (47)$$

is a solution, and φ satisfies the conditions

$$y = y_1, y_2 : \varphi = 0 \qquad (2)$$

if and only if

$$c = u(y_1) = u(y_2) \qquad (48)$$

In fact, if $(u'/T)'_{u=c} \neq 0$, this is the only regular solution, provided, of course, $u(y_1) = u(y_2)$, because from equation (34), c cannot be real and lie outside the range of u (since then $g(y)$ would be nonsingular). Furthermore, c cannot be within the range of u if the solution is to be regular, and then the boundary condition $\varphi_c = 0$ excludes one of the two solutions $\varphi_1(y)$ and $\varphi_2(y)$.

(8) SUFFICIENT CONDITIONS FOR THE EXISTENCE OF SELF-EXCITED AND NEUTRAL SUBSONIC MODES

Once the global necessary condition in the form

$$-\frac{\left\{\dfrac{u'}{T-M_1^2(u-c)^2}\right\}'}{u(y)-u(y_s)} \geqslant 0, \quad y_1 < y_s < y_2$$

holds, then we turn to the question of sufficient conditions for the existence of self-excited and neutral subsonic modes. Toward this we first exhibit the variational character of the inviscid problem for monotonic symmetrical type of mean flow, and then use the Sturm–Liouville theorem to show that there exists a neutrally-stable characteristic solution with $c = c_r = u_s$, and then construct unstable solutions that are contigous with the neutral modes, i.e. $c \to u_s$ as $c_i \to 0^+$.

THEOREM: *Regular neutral subsonic modes, i.e.*

$$\varphi = \varphi_s, \quad \alpha = \alpha_s, \quad c = c_s = u_s \tag{49}$$

are possible if the mean-flow profiles are monotonic, and if there is satisfied in the flow field

$$\left\{\frac{u'}{T-M_1^2(u-c)^2}\right\}'_{y=y_s} = 0$$

Proof: The proof proceeds on the lines of the one given by Friedrichs (1942) for the case of an incompressible fluid. First write equation (1) as

$$[N(y)\varphi']' + K(y)\varphi + \frac{\lambda}{T}\varphi = 0 \tag{50}$$

where

$$N(y) \equiv \frac{1}{T-M_1^2(u-c)^2}, \quad \lambda \equiv -\alpha^2$$

$$K(y) \equiv -\frac{\left\{\dfrac{u'}{T-M_1^2(u-c)^2}\right\}'}{u-c_s}$$

with the boundary conditions (2).

Let there be a point $y = y_s$ such that

$$\left\{\frac{u'}{T-M_1^2(u-c)^2}\right\}'_{y=y_s} = 0$$

and

$$\left\{\frac{u'}{T - M_1^2(u-c)^2}\right\}_{y=y_s} \neq 0$$

In addition let $u(y) > 0$ for $y_1 < y < y_2$ and assume that $K(y)$ is a regular function for $y_1 < y < y_2$. Then equation (50) is a real nonsingular equation which makes up a Sturm–Liouville problem with boundary condition (2). This characteristic-value problem possesses an infinite sequence of characteristic values. In order to show that there is a characteristic value α for which a neutral mode exists with $c = u(y_s)$, where $c = c_r$, we have to identify λ with $-\alpha^2$, i.e. we must show that at least one of the characteristic values of equation (50) is negative.

Multiply equation (50) by φ^*, and integrate from y_1 to y_2, and using the boundary conditions (2), one obtains

$$\lambda = \frac{\displaystyle\int_{y_1}^{y_2} [N|\varphi'|^2 - K|\varphi|^2]\,\mathrm{d}y}{\displaystyle\int_{y_1}^{y_2} \frac{|\varphi|^2}{T}\,\mathrm{d}y} \tag{51}$$

and as we shall presently see this is a characteristic-value problem for the parameter $\lambda = -\alpha^2$, and is associated with the variational principle

$$\delta \int_{y_1}^{y_2} \left[\frac{|\varphi'|^2}{T - M_1^2(u-c)^2} + \left\{ \frac{\left[\dfrac{u'}{T - M_1^2(u-c)^2}\right]'}{u - c_s} \right\} |\varphi|^2 \right] \mathrm{d}y = 0$$

subject to

$$\int_{y_1}^{y_2} \frac{|\varphi|^2}{T}\,\mathrm{d}y = \text{const.} \tag{52}$$

We now show that the variational principle associated with equation (50) gives the least characteristic value (51). In particular, if φ really satisfies equation (50) as well as the boundary conditions (2), then λ is determined from equation (51) by noting that λ is stationary if φ is given a slight variation $\delta\varphi$ satisfying only the boundary conditions (2). If $\delta\lambda$ corresponds to $\delta\varphi$, then

$$\delta\lambda = \frac{2I \displaystyle\int_{y_1}^{y_2} (N\varphi'\delta\varphi' - K\varphi\delta\varphi)\,\mathrm{d}y - 2J \int_{y_1}^{y_2} \frac{\varphi}{T}\,\delta\varphi\,\mathrm{d}y}{I^2} \tag{53}$$

where

$$I \equiv \int_{y_1}^{y_2} \frac{|\varphi|^2}{T} \, dy, \quad J \equiv \int_{y_1}^{y_2} [N|\varphi'|^2 - K|\varphi|^2] \, dy$$

Noting,

$$\int_{y_1}^{y_2} N\varphi'\delta\varphi' \, dy = -\int_{y_1}^{y_2} (N\varphi')'\delta\varphi \, dy$$

since

$$\delta\varphi' = (\delta\varphi)'$$

equation (53) becomes

$$-\frac{I}{2}\delta\lambda = \int_{y_1}^{y_2} \left[(N\varphi')' + K\varphi + \frac{\lambda}{T}\varphi\right] \delta\varphi \, dy \tag{54}$$

Therefore, from equation (50) one finds

$$\delta\lambda = 0 \tag{55}$$

Conversely, if $\delta\lambda$ vanishes for all variations $\delta\varphi$ satisfying the boundary conditions, and possessing a derivative which, like φ, is square integrable, then equation (54) shows that the corresponding function φ satisfies equation (50). Equation (55) merely shows that the characteristic value λ is stationary; in order to show that λ is a minimum, note that when $c_i \neq 0$, equation (18) gives

$$-\int_{y_1}^{y_2} \left[\left\{\frac{|\varphi'|^2}{T - M_1^2(u-c)^2}\right\} - \lambda_1 \frac{|\varphi|^2}{T}\right] dy$$

$$= \int_{y_1}^{y_2} \frac{\frac{d}{dy}\left\{\frac{u'}{T - M_1^2(u-c)^2}\right\}}{u-c} |\varphi|^2 \, dy$$

$$= \int_{y_1}^{y_2} \frac{(u-c_s)}{|u-c|^2} \frac{d}{dy}\left\{\frac{u'}{T - M_1^2(u-c)^2}\right\} |\varphi|^2 \, dy$$

Thus,

$$\int_{y_1}^{y_2} \left[\left\{\frac{|\varphi'|^2}{T - M_1^2(u-c)^2}\right\} - \lambda_1 \frac{|\varphi|^2}{T}\right] dy$$

$$= \int_{y_1}^{y_2} \frac{(u-c_s)^2}{|u-c|^2} K|\varphi|^2 \, dy < \int_{y_1}^{y_2} K|\varphi|^2 \, dy$$

From which,

$$-\lambda_1 < \frac{\displaystyle\int_{y_1}^{y_2}\left[-\left\{\frac{|\varphi'|^2}{T-M_1^2(u-c)^2}\right\}+K|\varphi|^2\right]\mathrm{d}y}{\displaystyle\int_{y_1}^{y_2}\frac{|\varphi|^2}{T}\,\mathrm{d}y}$$

or

$$-\lambda_1 < -\lambda$$

or

$$\lambda < \lambda_1$$

Thus,

$$\lambda = \min(\lambda_1)$$

where

$$\lambda_1 = \frac{\displaystyle\int_{y_1}^{y_2}\left[\left\{\frac{|\chi'|^2}{T-M_1^2(u-c)^2}\right\}+\frac{\left\{\dfrac{u'}{T-M_1^2(u-c)^2}\right\}'}{u-c}|\chi|^2\right]\mathrm{d}y}{\displaystyle\int_{y_1}^{y_2}\frac{|\chi|^2}{T}\,\mathrm{d}y} \tag{56}$$

where χ is any function possessing a square-integrable derivative and satisfying the boundary conditions (2).

Replace χ by the test function u, and recalling that $K(y) > 0$, one obtains

$$I_1\,\lambda = \int_{y_1}^{y_2}\frac{\left\{\dfrac{u'}{T-M_1^2(u-c)^2}\right\}' uu_s}{u-u_s}\,\mathrm{d}y$$

where

$$I_1 \equiv \int_{y_1}^{y_2}\frac{u^2}{T}\,\mathrm{d}y \neq I$$

or

$$I_1\,\lambda = -\int_{y_1}^{y_2} Kuu_s\,\mathrm{d}y < 0 \tag{57}$$

provided $u(y) > 0$. Therefore, λ is negative, and α real and positive, for which there is a neutrally-stable subsonic mode.

Next we deduce the existence of a self-excited mode in the neighbourhood of the neutral disturbance, i.e. unstable modes whose limit as $c_i \to 0^+$ is the neutral mode.

If

$$\frac{\mathrm{d}}{\mathrm{d}y}\left\{\frac{u'}{T-M_1^2(u-c)^2}\right\}_{u=c=c_r}=0 \tag{20}$$

holds somewhere within the flow field, one ensures the existence of neutral disturbances. If the conditions at such a point are denoted by a subscript s, one has a solution $c_s = u$, (where c_s is real).

Consider the characteristic-value problem of equation (1) with the parameter $-\alpha^2 = \lambda$. We have the global necessary condition in the form

$$-\frac{1}{u-c}\left\{\frac{u'}{T-M_1^2(u-c)^2}\right\}'>0$$

and let

$$u'(y)>0$$

Under these conditions, we have seen from equation (53) that the characteristic values λ_s are real and negative, and therefore α_s are real and positive. Also, the corresponding characteristic function φ_s is real.

That the neutral solution has a neighbouring self-excited solution will be shown here by considering the dependence of c on $\lambda = -\alpha^2$ in the characteristic-value problem of equation (1). Recalling that there is a characteristic function φ_s corresponding to the characteristic value $c = c_s$, $-\alpha_s^2 = \lambda_s$, one writes

$$\left[\frac{\varphi'}{T-M_1^2(u-c)^2}\right]'_{c=c_s}-\left[\frac{\left\{\dfrac{u'}{T-M_1^2(u-c)^2}\right\}'_{c=c_s}}{u-c_s}\right]\varphi_s+$$

$$+\frac{\lambda_s}{T}\varphi_s=0 \tag{58}$$

Multiply equation (1) by φ_s, equation (58) by φ and subtract, so that for the case $M_1 \ll 1$, one obtains

$$\frac{\mathrm{d}}{\mathrm{d}y}\left[\frac{\varphi_s\varphi'}{T-M_1^2(u-c)^2}-\frac{\varphi_s'\varphi}{T-M_1^2(u-c)^2}\right]+(\lambda-\lambda_s)\frac{\varphi\varphi_s}{T}+$$

$$-\left[\frac{\left\{\dfrac{u'}{T-M_1^2(u-c)^2}\right\}'}{u-c}-\frac{\left\{\dfrac{u'}{T-M_1^2(u-c_s)^2}\right\}'}{u-c_s}\right]\varphi\varphi_s=0 \tag{59}$$

Integrating between (y_1, y_2), one obtains

$$(\lambda - \lambda_s) \int_{y_1}^{y_2} \frac{\varphi\varphi_s}{T}\, dy$$

$$= \int_{y_1}^{y_2} \left[\frac{\left\{\dfrac{u'}{T - M_1^2(u-c)^2}\right\}'}{u-c} - \frac{\left\{\dfrac{u'}{T - M_1^2(u-c_s)^2}\right\}'}{u-c_s} \right] \varphi\varphi_s\, dy$$

which in the limit $\lambda \to \lambda_s$, $c \to c_s$, $\varphi \to \varphi_s$ gives

$$-\left(\frac{d\lambda}{dc}\right)_{\alpha=\alpha_s} \int_{y_1}^{y_2} \frac{|\varphi_s|^2}{T}\, dy$$

$$= \underset{c \to c_s}{\mathrm{Lt}} \int_{y_1}^{y_2} dy \left[\frac{\left\{\dfrac{u'}{T - M_1^2(u-c)^2}\right\}'_{c=c_s}}{u-c_s} \right] \frac{|\varphi_s|^2}{u-c} \tag{60}$$

The right-hand side of equation (60) is

$$\int_{y_1}^{y_2} \frac{|\varphi_s|^2}{u-c} \left[- \frac{\left\{\dfrac{u'}{T - M_1^2(u-c)^2}\right\}'_{c=c_s}}{u-c_s} \right] dy$$

$$= \int_{y_1}^{y_2} \frac{\left[- \dfrac{\left\{\dfrac{u'}{T - M_1^2(u-c)^2}\right\}'_{c=c_s}}{u-c_s} \right] (u-c_r)}{|u-c|^2} |\varphi_s|^2\, dy +$$

$$+ i \int_{y_1}^{y_2} \frac{\left[- \dfrac{\left\{\dfrac{u'}{T - M_1^2(u-c)^2}\right\}_{c=c_s}}{u-c_s} \right] c_i}{|u-c|^2} |\varphi_s|^2\, dy$$

The limit of the reaı part becomes the principal value of the integral

$$\int_{y_1}^{y_2} \frac{\left[\dfrac{\left\{\dfrac{u'}{T - M_1^2(u-c)^2}\right\}'_{c=c_s}}{u-c_s} \right]}{(u-c_s)} |\varphi_s|^2\, dy$$

The imaginary part tends to the limit

$$\pi \left[-\frac{\left\{ \dfrac{u'}{T - M_1^2 (u-c)^2} \right\}'_{c=c_s}}{u_s'(u - c_s)} \right] |\varphi_s|^2 > 0$$

as $c_i \to 0^+$. Equation (60) gives then an equation of the form

$$-\frac{d\lambda}{dc} = A + iB, \quad B > 0$$

of

$$-\frac{dc}{d\lambda} = \frac{dc}{d(\alpha^2)} = \frac{A - iB}{A^2 + B^2} \tag{61}$$

If α^2 decreases slightly, equation (61) shows that c_i becomes positive, which proves the existence of a self-excited disturbance in the neighbourhood of the neutral disturbance. In other words, when viscosity and heat conductivity are neglected, if the flow can execute a small neutral subsonic oscillation of finite wavelength, it can also execute self-excited subsonic oscillation of longer wavelengths, and damped oscillation of shorter wavelengths.

(9) THE NUMBER OF UNSTABLE MODES

Let the global necessary condition for the existence of nonneutral or neutral subsonic modes in the form

$$K(y) = -\frac{\left\{ \dfrac{u'}{T - M_1^2 (u-c)^2} \right\}'}{u - c_s} \geqslant 0$$

hold, where c_s is real. Assume that $(u - c_s)$ has one (and only one) zero in (y_1, y_2), otherwise the flow would be stable.

Consider the characteristic-value problem

$$[N(y)\chi']' + K(y)\chi + \frac{\lambda}{T}\chi = 0 \tag{62}$$

$$y = y_1, y_2 : \chi = 0$$

as in (50). If we restrict ourselves to subsonic disturbances, this is an ordinary Sturm–Liouville problem, and possesses an increasing sequence $\lambda_1, \lambda_2, \ldots$ of characteristic values. λ_1 may be characterised, as we saw previously, as the minimum of the functional

$$\mathscr{F}\{\varphi\} = \int_{y_1}^{y_2} [N|\varphi'|^2 - K|\varphi|^2] \, dy \tag{63}$$

among continuously-differentiable functions satisfying the boundary conditions in (62), and normalised by

$$\int_{y_1}^{y_2} \frac{|\varphi|^2}{T}\,\mathrm{d}y = 1 \tag{64}$$

The minimum is attained for $\varphi = \chi_1$, the first characteristic function. Similarly, λ_n is the minimum for such φ as are also orthogonal to the first $(n-1)$ characteristic functions, and is attained for $\varphi = \chi_n$. Whenever the characteristic values λ_k are negative, one identifies $\lambda_k = -\alpha_{sk}^2$, which has a corresponding neutral characteristic function χ_k. According to the sufficiency conditions we just established, unstable modes are then expected to exist in the neighbourhood of the neutral modes, for $\alpha^2 < \alpha_{sk}^2$.

THEOREM: *For the type of mean-flow profiles that bear the global necessary condition to hold in the flow field, there can be no more than $(n-1)$ linearly-independent unstable characteristic functions of (62) if $\alpha^2 \geqslant -\lambda_n$. (Or the number of distinct unstable subsonic modes cannot exceed the number of neutral subsonic modes associated with the points where $(u'/T)'_{u=c} = 0$.)*

Proof: The proof proceeds on the lines given by Howard (1964) for the case of an incompressible fluid. Let φ_1, φ_2 be two (not necessarily different) unstable characteristic functions for equation (1) with the same characteristic value α^2 and values c_1, c_2 respectively (both have $c_i > 0$); then we have

$$\int_{y_1}^{y_2} \left[-\varphi_1^* \left\{ \frac{\varphi_2'}{T - M_1^2(u-c)^2} \right\}' + \frac{\alpha^2}{T}\,\varphi_1^*\varphi_2 \right] \mathrm{d}y$$

$$= -\int_{y_1}^{y_2} \frac{\left\{ \dfrac{u'}{T - M_1^2(u-c)^2} \right\}'}{(u-c_2)}\,\varphi_1^*\varphi_2\,\mathrm{d}y \tag{65}$$

and similarly

$$\int_{y_1}^{y_2} \left[-\varphi_2 \left\{ \frac{\varphi_1^{*\prime}}{T - M_1^2(u-c)^2} \right\}' + \frac{\alpha^2}{T}\,\varphi_2\varphi_1^* \right] \mathrm{d}y$$

$$= -\int_{y_1}^{y_2} \frac{\left\{ \dfrac{u'}{T - M_1^2(u-c)^2} \right\}'}{(u-c_1^*)}\,\varphi_2\varphi_1^*\,\mathrm{d}y \tag{66}$$

where for the case $M_1 \ll 1$, the variation in c is negligible in the braces to first approximation. Set

$$\frac{\varphi_k}{u - c_k} \equiv F_k \tag{67}$$

so that from (65) and (66) one obtains

$$0 = \int_{y_1}^{y_2} \left[-\varphi_1^* \varphi_2 \left\{ \frac{u'}{T - M_1^2(u-c)^2} \right\}' \right]$$

$$\times \left[\frac{1}{u - c_2} - \frac{1}{u - c_1^*} \right] dy$$

or

$$0 = \int_{y_1}^{y_2} K F_1^* F_2 (u - c_s)(c_2 - c_1^*)\, dy$$

or

$$\int_{y_1}^{y_2} K(u - c_s) F_1^* F_2 \, dy = 0 \tag{68}$$

Suppose that $\varphi_1, \varphi_2, \ldots, \varphi_n$ are n linearly-independent unstable characteristic functions of equation (1) for the same value of α and values $c_1, c_2, \ldots, c_n$ of c. Since these functions are linearly independent, one constructs

$$\psi = \sum_{k=1}^{n} a_k \varphi_k \tag{69}$$

which is orthogonal to the first $(n-1)$ characteristic functions $\chi_1, \chi_2, \ldots, \chi_{n-1}$ of the Sturm–Liouville problem (1); let

$$\int_{y_1}^{y_2} \frac{|\psi|^2}{T} \, dy = 1 \tag{70}$$

Now

$$\left\{ \frac{\psi'}{T - M_1^2(u-c)^2} \right\}' - \frac{\alpha^2}{T} \psi = \sum_{k=1}^{n} a_k \left\{ \frac{u'}{T - M_1^2(u-c)^2} \right\}' \frac{\varphi_k}{u - c_k}$$

$$= - \sum_{k=1}^{n} a_k F_k K (u - c_s)$$

or

$$\left\{ \frac{\psi'}{T - M_1^2(u-c)^2} \right\}' + K\psi = \frac{\alpha^2}{T} \psi + \sum_{k=1}^{n} a_k F_k (c_s - c_k) K \tag{71}$$

Multiply this by ψ^*, and integrate over the flow domain; then one obtains

$$\int_{y_1}^{y_2}\left[-\left\{\frac{|\psi'|^2}{T-M_1^2(u-c)^2}\right\}+K|\psi|^2\right]\mathrm{d}y$$

$$=\alpha^2\int_{y_1}^{y_2}\frac{|\psi|^2}{T}\,\mathrm{d}y+\int_{y_1}^{y_2}K\sum_{k=1}^{n}\sum_{l=1}^{n}a_l^*F_l^*(u-c_l)a_kF_k(c_s-c_k)$$

$$=\alpha^2+\int_{y_1}^{y_2}K\sum_{k=1}^{n}\sum_{l=1}^{n}a_l^*F_l^*(c_s-c_l)a_kF_k(c_s-c_k)\,\mathrm{d}y+$$

$$+\sum_{k=1}^{n}\sum_{l=1}^{n}\int_{y_1}^{y_2}K(u-c_s)a_l^*\,F_l^*a_kF_k(c_s-c_k)\,\mathrm{d}y$$

or

$$-\alpha^2=\int_{y_1}^{y_2}\left[\left\{\frac{|\psi'|^2}{T-M_1^2(u-c)^2}\right\}-K|\psi|^2\right]\mathrm{d}y+$$

$$+\int_{y_1}^{y_2}K\left|\sum_{k=1}^{n}a_k(c_s-c_k)F_k\right|^2\mathrm{d}y\geqslant\mathscr{F}\{\psi\}$$

or

$$-\alpha^2\geqslant\lambda_n \tag{72}$$

Therefore, if there are to be as many as n linearly-independent unstable characteristic functions of (1) one must have $\alpha^2\leqslant-\lambda_n$. Indeed, one must have $\alpha^2<-\lambda_n$, since if the equality is to hold, then

$$\sum_{k=1}^{n}a_k(c_s-c_k)F_k=0$$

so that

$$\sum_{k=1}^{n}a_k(c_s-c_k)\left[\left\{\frac{\varphi_k'}{T-M_1^2(u-c)^2}\right\}-\frac{\alpha^2}{T}\varphi_k\right]$$

$$=\left\{\frac{u'}{T-M_1^2(u-c)^2}\right\}'\sum_{k=1}^{n}a_k(c_s-c_k)F_k=0 \tag{73}$$

Putting

$$g=\sum_{k=1}^{n}a_k(c_s-c_k)\varphi_k \tag{74}$$

one finds

$$\int_{y_1}^{y_2} \left[\left\{ \frac{|g'|^2}{T - M_1^2(u-c)^2} \right\} + \frac{\alpha^2}{T} |g|^2 \right] dy = 0 \tag{75}$$

so that for subsonic modes,

$$g \equiv 0 \tag{76}$$

Since $c_s - c_k \neq 0$ this implies the linear dependence of $\varphi_1, \varphi_2, \ldots, \varphi_n$, and hence the contradiction.

Thus if α^2 is larger than the absolute value of the lowest characteristic value $-\lambda_1$, the flow is stable. And eventually λ_n becomes positive, and thereafter the relevance to the stability problem no longer exists. Indeed, any conditions ensuring that the Sturm–Liouville problem (62) has only positive characteristic values are sufficient conditions for stability.

7

Subsonic and Supersonic Disturbances

INTRODUCTION

The previous chapter showed that whereas a well-established inviscid theory exists for subsonic disturbances, no such theory exists for supersonic disturbances. Here we shall make discussions of the nature of supersonic disturbances as distinct from that of the subsonic disturbances. First a phase-plane analysis is made from which results show that whereas a given basic flow executes a unique set of subsonic oscillations, no uniqueness exists in executing supersonic oscillations. Next, an investigation is made of the stability characteristics of the surface waves generaged in a liquid layer adjacent to a high-speed gas stream. It turns out that the nature of these surface waves depends on the type of disturbances present in the high-speed gas stream. For the case of subsonic disturbances, it is found that the energy transfer from the gas stream to the surface waves is contributed predominantly by the Fourier component of the normal gas-pressure force field in phase with the slope of the wavy surface. For the case of supersonic disturbances, this energy transfer is contributed predominantly by the Fourier component of the gas-pressure field in phase with the surface-wave displacement and is related to the presence of travelling periodic waves in the gas stream – this energy transfer is always found to promote the growth of the surface waves.

(1) INVISCID PRESSURE DISTURBANCES

The inviscid problem is governed by (see Chapter IV)

$$\mathrm{i}(u-c)\gamma+\rho(\varphi'+\mathrm{i}f)+\rho'\varphi=0 \tag{1}$$

$$\rho[\mathrm{i}(u-c)f+u'\varphi]=-\frac{\mathrm{i}\pi}{\sigma M_1^2} \tag{2}$$

$$\alpha^2\rho[\mathrm{i}(u-c)\varphi]=-\frac{\pi'}{\sigma M_1^2} \tag{3}$$

$$\rho[\mathrm{i}(u-c)\theta+T'\varphi]=-\rho T(\sigma-1)(\varphi'+\mathrm{i}f) \tag{4}$$

$$\frac{\pi}{p}=\frac{\gamma}{\rho}+\frac{\theta}{T} \tag{5}$$

Using equation (5), equation (1) becomes

$$\varphi'+\mathrm{i}f-\frac{T'}{T}\varphi+\mathrm{i}(u-c)\left(\frac{\pi}{\rho T}-\frac{\theta}{T}\right)=0 \tag{6}$$

Using (6), equation (4) becomes

$$\rho[\mathrm{i}(u-c)\theta+T'\varphi]=\frac{\mathrm{i}\pi}{\sigma}(\sigma-1)(u-c) \qquad (7) \qquad (^7)$$

Using (7), equation (3) gives

$$\frac{\mathrm{i}(u-c)}{T}\theta=\frac{\mathrm{i}}{\sigma\rho T}(\sigma-1)(u-c)\pi+\frac{T'}{T}\left[-\frac{\mathrm{i}\pi'}{\rho\alpha^2\sigma M_1^2(u-c)}\right] \tag{8}$$

From (6) and (7), one obtains

$$\mathrm{i}f=\frac{1}{u-c}\left[-\frac{\mathrm{i}\pi}{\rho\sigma M_1^2}-u'\left\{\frac{\mathrm{i}\pi'}{\rho\alpha^2\sigma M_1^2(u-c)}\right\}\right] \tag{9}$$

Using equations (3), (8) and (9), (6) gives

$$\left[\frac{\mathrm{i}\pi'}{\sigma M_1^2}\,\frac{1}{\rho\alpha^2(u-c)}\right]'+\frac{1}{u-c}\left[-\frac{\mathrm{i}\pi}{\rho\sigma M_1^2}-\frac{\mathrm{i}u'\pi'}{\rho\alpha^2\sigma M_1^2(u-c)}\right]+$$

$$+\,\mathrm{i}(u-c)\;\frac{\pi}{\rho T}-\frac{\mathrm{i}(\sigma-1)(u-c)}{\sigma\rho T}\pi+\frac{\mathrm{i}T'\pi'}{\rho\alpha^2\sigma M_1^2T(u-c)}=0$$

or

$$\pi''+\left(\frac{T'}{T}-\frac{2u'}{u-c}\right)\pi'-\alpha^2\left[1-\frac{M_1^2(u-c)^2}{T}\right]\pi=0 \tag{10}$$

Equation (10) was numerically integrated by Mack (1965) for a laminar boundary layer along an insulated plate at various Mach numbers. Mack's (1965) results (shown in Fig. 1) showed that there is only one mode at low Mach numbers, and more modes appear at high Mach numbers.

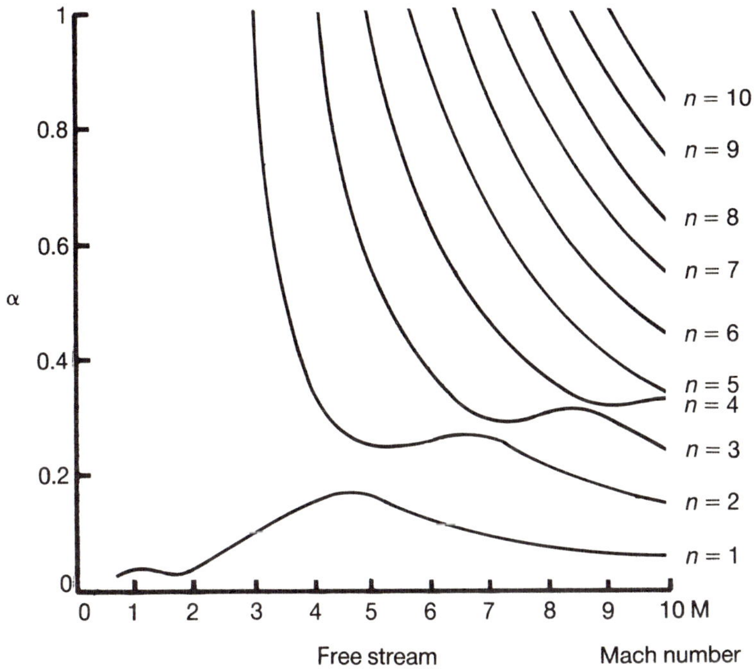

Fig. 1. Wave numbers of inviscid neutral solutions (insulated wall-boundary layer). From Mack (1965) as given in R. Betchov and W.O. Criminale, *Stability of Parallel Flows*, Academic Press, 1967.

(2) A PHASE-PLANE ANALYSIS

A phase-plane analysis of the system represented by equation (1) provides an illustration of the behaviour of the disturbances governed by equation (10). In particular, it reveals:

(i) the conditions for boundedness at infinity for disturbances;
(ii) the dependence of the spatial distribution of a disturbance on the boundary condition;
(iii) how the spatial growth or decay of disturbances depends upon the mean-flow parameters.

Referring to equation (10), if c is complex then so is the solution, and so it is better to consider neutral disturbances, and then π is real. In order to represent the behaviour of the disturbance in a phase plane, one writes equation (10) as

$$\pi' = g$$

$$g' = \left(-\frac{T'}{T} + \frac{2u'}{u-c}\right) g + \alpha^2 \left[1 - \frac{M_1^2(u-c)^2}{T}\right]\pi \tag{11}$$

One notices that this is an autonomous system. To every (π, g) of the system (11) in physical space, there corresponds a point in the phase plane. Conversely, to every point in the phase plane where the right-hand sides of equation (11) are both nonzero there corresponds one and only one value (π, g) of the given system. To a continual spatial distribution of the disturbance, there corresponds a motion of the representative point in the phase plane.

The equilibrium distribution of the system corresponds to those points of the phase plane where both the right-hand sides of equation (11) vanish simultaneously. One immediately notices that

$$\pi = 0, \quad g = 0$$

is the singular point of the system (11). At such points, Cauchy's conditions for the uniqueness of the solution of the differential equation (1) are not satisfied (for details, see Butenin (1965)).

Write equation (11) as

$$\begin{pmatrix} \pi' \\ g' \end{pmatrix} = \begin{pmatrix} 0 & 1 \\ \alpha^2 \left[1 - \dfrac{M_1^2(u-c)^2}{T}\right] & -\dfrac{T'}{T} + \dfrac{2u'}{u-c} \end{pmatrix} \begin{pmatrix} \pi \\ g \end{pmatrix} \tag{12}$$

It is of interest to study the behaviour of the solution in the neighbourhood of the singular point, since the nature of the singular point determines the character of the undisturbed mean flow. And the nature of the solutions in the neighbourhood of the singular point depends strongly on the characteristic values of the matrix coefficient in equation (12).

We consider the cases for which

$$\left[1 - \frac{M_1^2(u-c)^2}{T}\right] \neq 0$$

i.e. we exclude from the consideration neutral sonic disturbances, since in this case the set of singular points is infinite and fills up the line

$$\pi = \frac{\dfrac{2u'}{u-c} - \dfrac{T'}{T}}{\alpha^2 \left[1 - \dfrac{M_1^2(u-c)^2}{T}\right]} g$$

The characteristic values of the matrix coefficient in equation (12) are given by

$$\begin{vmatrix} -\lambda & 1 \\ \alpha^2 \left[1 - \dfrac{M_1^2(u-c)^2}{T}\right] & -\dfrac{T'}{T} + \dfrac{2u'}{u-c} - \lambda \end{vmatrix} = 0 \tag{13}$$

From which

$$\lambda^2 + \left(\frac{T'}{T} - \frac{2u'}{u-c}\right)\lambda - \alpha^2\left[1 - \frac{M_1^2(u-c)^2}{T}\right] = 0$$

or

$$\lambda = \left(-\frac{T'}{T} + \frac{2u'}{u-c}\right) \pm \left[\left(-\frac{T'}{T} + \frac{2u'}{u-c}\right)^2 + \right.$$

$$\left. + \alpha^2\left\{1 - \frac{M_1^2(u-c)^2}{T}\right\}\right]^{1/2} \qquad (14)$$

For subsonic disturbances, equation (14) gives

$$\lambda_1 > 0, \quad \lambda_2 < 0$$

The singular point is then of saddle type. The integral curves are a family of hyperbolas with a pair of asymptotes (see Fig. 2)

$$g = \lambda_1 \pi$$

$$g = \lambda_2 \pi$$

The character and configuration of the trajectories and the direction of motion of the representative point along them enable one to see the behaviour of the disturbances. The pair of asymptotes splits the phase plane into four regions, and any solution that starts in one of these must stay in it, otherwise two solutions would intersect. And two solutions through the same point must be identical unless that point is the singular point. One sees that for arbitrary boundary conditions except for those which correspond to the asymptotes

$$g = \lambda_1 \pi, \quad \text{for } y < 0$$

$$g = \lambda_2 \pi, \quad \text{for } y > 0$$

the representative point will move away from the singular point (i.e. it grows and does so exponentially as $y \to \pm\infty$). On the other hand, a solution that starts on these asymptotes decays with distance y. However, a representative point moves along these aysmptotes towards the origin, but does not reach it in a finite distance y. And it is precisely this kind of spatial behaviour that one would be after in trying to obtain a description of the disturbances, and therefore only that set of subsonic disturbances with a representative point on these asymptotes is the one that is acceptable. And this is so also, for any other arbitrary trajectory, when a representative point moves away from the singular point, the disturbance gradients become arbitrarily large so that the viscous effects necessarily enter the picture and one can no longer treat the disturbances as being inviscid.

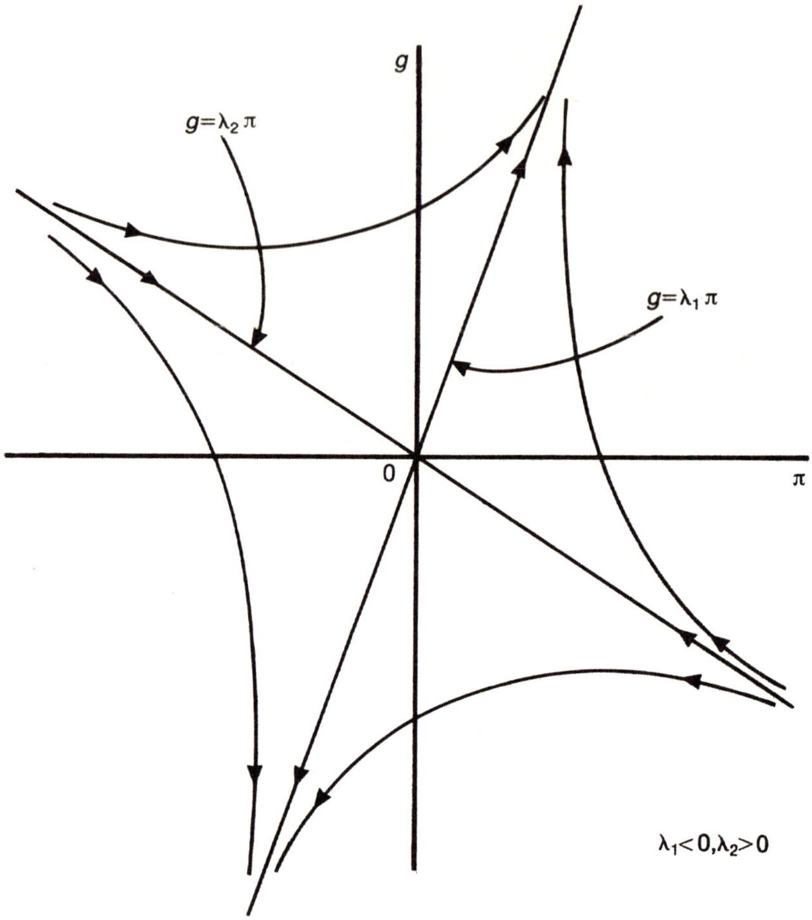

Fig. 2

This means that a given mean flow executes a unique set of subsonic oscillations, which are given by

$$\frac{\pi'}{\pi} = \lambda_2, \quad \text{for } y > 0$$

$$\frac{\pi'}{\pi} = \lambda_1, \quad \text{for } y < 0 \tag{15}$$

On the other hand, for supersonic disturbances, and for

$$\left(-\frac{T'}{T} + \frac{2u'}{u-c}\right)^2 > \alpha^2 \left[-1 + \frac{M_1^2(u-c)^2}{T}\right]$$

both the characteristic values λ_1, λ_2 are real, and of the same sign, and the singular point is a node (see Fig. 3). One has

(i) $\begin{matrix} \lambda_1 > 0 \\ \lambda_2 > 0 \end{matrix}$ if $\left(\dfrac{T'}{2T} + \dfrac{u'}{u-c}\right) > 0$

(ii) $\begin{matrix} \lambda_1 < 0 \\ \lambda_2 < 0 \end{matrix}$ if $\left(-\dfrac{T'}{2T} + \dfrac{u'}{u-c}\right) < 0$

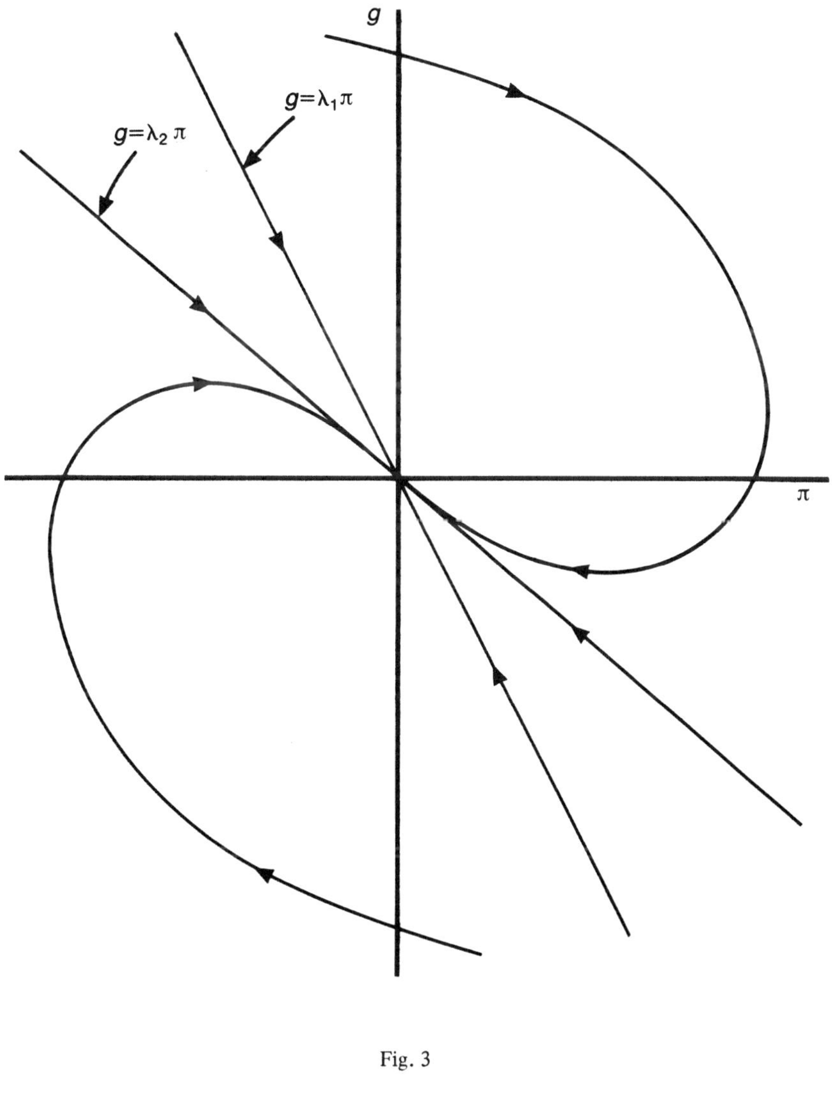

Fig. 3

It is clear that such supersonic oscillations cannot be found uniquely for a given mean-flow condition.

$$\left(-\frac{T'}{T}+\frac{2u'}{u-c}\right)^2<\alpha^2\left[-1+\frac{M_1^2(u-c)^2}{T}\right]$$

the characteristic values are complex conjugates of each other, and the singular point is a spiral point. The solutions in the neighbourhood of the singular point are spirals which wind around the origin (see Fig. 4). Again it is clear that such supersonic oscillations cannot be found uniquely for a given mean-flow condition.

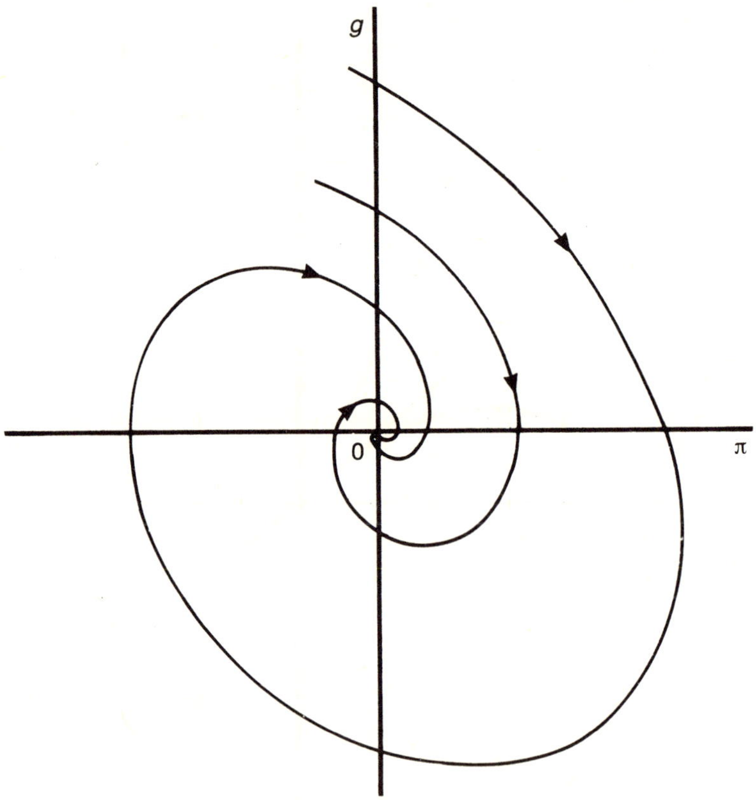

Fig. 4

It is of interest to note that the nonuniqueness of the supersonic oscillations is also indicated by a consideration of the characteristic-value problem associated with equation (10). In the far field where the mean-flow gradients vanish, equation (10) becomes

$$\pi''-\frac{\alpha^2}{T_1}\,[T_1-M_1^2(V_1-c)^2]\pi=0 \tag{16}$$

from which

$$y \to \infty : \pi \sim e^{-s_1 y} \tag{17}$$

where

$$s_1 = \alpha \sqrt{\frac{1}{T_1} [T_1 - M_1^2 (V_1 - c)^2]}$$

Now, corresponding to neutral supersonic disturbances, s_1 turns out to be purely imaginary so that the homogeneous boundary condition

$$y \to \infty : \pi' + s_1 \pi = 0$$

expressing boundedness at infinity (that one could construct for the case of subsonic disturbances) no longer arises. As a result, for this case, a well-posed characteristic-value problem does not exist so that the characteristic values are continuous and not discrete; i.e. a solution satisfying the boundary conditions can always be found for arbitrary values of c, α.

(3) GENERATION OF SURFACE WAVES IN A LIQUID LAYER ADJACENT TO A GAS STREAM

Early studies of the physical processes through which a wind blowing over a water surface generates surface waves in the water were made by Kelvin (1887). Kelvin considered an inviscid model that assumed a tengential discontinuity between uniform flows in air and water, and predicted a critical wind speed for the onset of surface waves. Miles (1957) considered an extension of Kelvin's model to allow for variation of wind speed with elevation, thereby posing a problem in hydrodynamic stability for flows over a nonrigid boundary. The flexibility of the boundary enables it to receive energy from the shear flow. The dynamic instability present in Miles' mechanism is different from the static instability present in Kelvin's mechanism. The latter is merely a consequence of an aerodynamic pressure that is in phase with the displacement of the wave and greater than the sum of the gravitational and capillary forces resisting this displacement. Dynamic instability, on the other hand, is associated with an aerodynamic pressure that is in phase with the slope of the surface wave. Here, we consider an extension of Miles' model to take into account the compressibility effects in the wind (Shivamoggi, 1978a).

We consider a simple wave train that is stationary relative to the surface waves in the liquid layer superposed on the incoming gas stream, and then investigate the conditions under which the reaction of the disturbed flow is just sufficient to sustain the surface waves. The justification of the inviscid model for the determination of a first approximation to the disturbed motion of the gas stream and the consequent energy transfer to the surface waves in the liquid layer comes from the fact that the predominant force field at the interface is due to the pressure field of the gas stream, which is not affected by the presence of the viscous boundary layer and which is considerably larger than the viscous frictional force at the interface.

It turns out that the nature of the surface waves generated in the liquid layer depends significantly on the type of disturbances present in the gas streatm. For case of subsonic disturbances, it is found that the energy transfer from the gas stream to the surface waves is contributed predominantly by the Fourier component of the normal gas-pressure force field in phase with the slope of the wavy surface. This appears to be similar to the so-called 'sheltering' mechanism by Jefreys (1924, 1925) in the context of generation of surface waves in water by a wind blowing over it, which is based on a flow separation from the leeward side of each wave on the surface so that there is a pressure difference between the windward side and the leeward side of each crest, and if the wind speed exceeds the wave speed, energy is thereby transferred from the air to the water. In calculating this energy transfer, use is made of the fact that the latter is determined by the conditions at the critical point $u(y) = c$, since it is known that the flexure of the interface causes the wavetrain in the flow direction with a speed c, and the critical point may then occur away from the interface. For the case of supersonic disturbances, this energy transfer is contributed predominantly by the Fourier component of the gas-pressure field in phase with the surface-wave displacement and is related to the presence of travelling periodic waves in the gas stream – this energy transfer is always found to promote the growth of the surface waves.

(i) Surface-wave generation by subsonic disturbances

Consider an initially-quiescent liquid of infinite depth in the region $y < 0$, and a gas stream in the x-direction past the liquid surface through the region $y > 0$. The gas stream is taken to be parallel, i.e. the flow properties are considered to vary negligibly over distances comparable with the wavelength, and in the direction of the surface waves in the liquid layer so that the disturbances due to the wavy interface given by

$$y_0 = a\,\mathrm{e}^{\mathrm{i}\alpha(x-ct)} \tag{18}$$

are two dimensional and strictly periodic in the x-coordinate. In the linearised theory for the disturbances that we are concerned with here, one ignores the flow separation from the wave crests and requires the surface-wave amplitude to be small in comparison with the wavelength and also that the surface curvature be sufficiently small. In the present case, the motion of the liquid below the interface is neglected.

The perturbations in the flow properties are considered to be of the form $\mathrm{e}^{\mathrm{i}\alpha(x-ct)}$. Under the approximations of parallel flows, the equations of continuity, motion, energy and state yield, on reduction (see Chapter IV),

$$\frac{\mathrm{d}}{\mathrm{d}y}\left[\frac{(u-c)\varphi' - u'\varphi}{T - M_1^2(u-c)^2}\right] - \frac{\alpha^2(u-c)}{T}\varphi = 0 \tag{19}$$

The boundary conditions are

$$y \to 0 : \varphi = 0$$

$$y \to \infty : \varphi \to 0 \tag{20}$$

From equation (3), one obtains

$$\pi(0) = -\int_0^\infty \pi' \, dy = \sigma M_1^2 \int_0^\infty \frac{i\alpha^2 (u-c)}{T} \varphi \, dy \tag{21}$$

which implies that the pressure distribution at the interface is governed by the cumulative action of the disturbance over the whole flow field, and is affected very little by the phenomena near the interface. Using equation (19), one obtains

$$p_s = \pi(0) = -\frac{i\sigma M_1^2 (u-c)}{T - M_1^2 (u-c)^2} \varphi' \tag{22}$$

Since the work done by the pressure forces on the wavy interface is

$$c \int p_s \frac{\partial y_0}{\partial x} \, dx$$

it follows that in the absence of viscosity, the rate of energy transfer from the gas stream to unit area of the wavy interface is proportion to $-\mathrm{Im}\{p_s\}$. The growth of the surface waves in the liquid layer is then supposed to occur when the energy supply exceeds the rate of energy dissipation by viscosity in the fluid below the interface.

Note from (22) that

$$-\mathrm{Im}\{p_s\} \approx \frac{\sigma M_1^2 c_i}{T - M_1^2 (u-c)^2} \mathrm{Im}\{\varphi'\} \tag{23}$$

Introduce the Wronskian

$$W \equiv \frac{i}{2} \frac{\varphi\varphi^{*\prime} - \varphi'\varphi^*}{T - M_1^2 (u-c)^2} \tag{24}$$

then one has (see equation (VI.27))

$$[W]_c \equiv W(y_c^+) - W(y_c^-) = \pi \frac{(u'/T)'_{u=c}}{u_c'} |\varphi_c|^2 \tag{25}$$

provided in this limit $c_i > 0$, $u_c^+ > u_c^-$.

Since $\varphi \to 0$ for $y \to \infty$, $W \equiv 0$ for $y > y_c$, so that one obtains

$$y < y_c : W(y) = -[W]_c \tag{26}$$

Noting that $\varphi = u - c$ is a solution of equation (19) in the limit $\alpha \to 0$, one finds for the neighbourhood of the interface

$$W(y) = \frac{i}{2} \frac{(u-c)}{T - M_1^2(u-c)^2} (\varphi^{*\prime} - \varphi')$$

or

$$W(y) = \frac{(u-c)}{T - M_1^2(u-c)^2} \operatorname{Im}\{\varphi'\} \tag{27}$$

From (23), (25)–(27) there follows, as $c_i \to 0$, that

$$-\operatorname{Im}\{p_s\} \approx -\pi\sigma M_1^2 \frac{(u'/T)'_{u=c}}{u'_c} |\varphi_c|^2 \tag{28}$$

which shows that, in the absence of viscosity, the energy transfer from the gas stream through the subsonic disturbances is positive (such as to promote the growth of the surface waves in the liquid layer) when

$$(u'/T)'_{u=c} < 0 \tag{29}$$

(ii) Surface-wave generation by supersonic disturbances

In this case the gas flowing past the liquid surface is assumed to have a uniform velocity V_1 and a uniform temperature T_1. Noting that for the case of supersonic disturbances, one has periodic waves, it follows from equation (22) that the pressure exerted by the gas at the interface is equal to

$$\Delta p_g(0) = \left[\frac{\rho_g \, i\sigma M_1^2 \, a\alpha(V_1 - c)}{\left\{ \dfrac{M_1^2(V_1 - c)^2}{T} - 1 \right\}^{1/2}} \right] e^{i\alpha(x-ct)} \tag{30}$$

The linearised equations of motion of the liquid are

$$\rho_l \frac{\partial \mathbf{V}_l}{\partial t} = -\nabla(\Delta p_l) \tag{31}$$

$$\nabla \cdot \mathbf{V}_l = 0 \tag{32}$$

where $\mathbf{V}_l$ is the liquid velocity. The boundary conditions are

(i) $y \to -\infty : |\mathbf{V}_l| \to 0$ (33)

(ii) $y = 0 : \dfrac{\partial y_0}{\partial t} = \mathbf{V}_l$ (34)

$$\Delta p_g = \Delta p_l \tag{35}$$

The solutions to equations (31) and (32), subject to (33) and (34), are

$$u_1 = a\alpha c\, e^{\alpha y}\, e^{i\alpha(x-ct)}$$

$$V_1 = -\, i a\alpha c\, e^{\alpha y}\, e^{i\alpha(x-ct)}$$

$$\Delta p_1 = \rho_1\, a\alpha c^2\, e^{\alpha y}\, e^{i\alpha(x-ct)} \tag{36}$$

Using (36), (35) gives

$$c^2 = \frac{\rho_g}{\rho_1} \frac{i\alpha M_1^2(V_1 - c)}{\left\{\dfrac{M_1^2(V_1 - c)^2}{T_1} - 1\right\}^{1/2}} \tag{37}$$

from which it is clear that for supersonic disturbances

$$\mathrm{Im}\,\{c\} > 0 \tag{38}$$

so that there always appears to be an energy transfer from the gas stream through the supersonic disturbances to the surface waves leading to their amplification – in significant contrast to the situation associated with the subsonic disturbances.

8

Asymptotic Analysis of Viscous Disturbances

INTRODUCTION

Because the situation of primary interest corresponds to large values of the Reynolds number, it is inevitable that the asymptotic methods of analysis adopted here below should play an important role in the theory. In the asymptotic theory, the sacrifice of the uniformity of approximation is compensated by greater simplicity. We shall pose a singular-perturbation problem for the linearised disturbances in a parallel gas flow associated with a thin viscous layer in the neighbourhood of the critical point $u = c$, and use the method of matched asymptotic expansions to develop a simplified system of equations (of the same accuracy) and the associated asymptotic matching conditions (Shivamoggi, 1978b). We will develop straightforward asymptotic expansions to represent the viscous disturbances which are valid everywhere except at the point $u = c$. The point $u = c$ is now in the viscous theory found to be a turning point associated with the mathematical problem. We will then develop uniformly-valid asymptotic expansions corresponding to a close neighbourhood of the turning point, and match them asymptotically to the straightforward asymptotic expansions. We will develop a convergent-series solution valid in a complex neighbourhood of the point $u = c$, and an asymptotic-series solution (WKB method) valid away from the point $u = c$. The results will be used to discuss some general characteristics of the curve of neutral stability. We will discuss the asymptotic properties of the various solutions; the accuracy of the simplified system of equations deduced by using the method of matched asymptotic expansions will be checked, and the proper path of integration associated with the WKB solution will be established by comparing it with the convergent-series solution in the asymptotic limit, and the associated physical interpretation will be indicated. Stokes' phenomenon will be indicated to arise at the point $u = c$; the configuration of Stokes' lines at the point $u = c$ will be indicated, and the associated physical interpretation and applications will be indicated. Next, uniformly-valid asymptotic solutions for viscous disturbances valid in a finite region containing the critical point $u = c$ will be developed using the method of multiple scales.

(1) VISCOUS PROBLEM

The governing differential equations for the linearised viscous disturbances are (see Chapter III)

$$\rho(u-c)\gamma + \rho(\varphi' + \mathrm{i}f) + \rho'\varphi = 0 \tag{1}$$

$$\alpha\rho\,[\mathrm{i}(u-c)f + u'\varphi] = -\frac{\mathrm{i}\alpha\pi}{\sigma M_1^2} + \frac{1}{R_L}\,[f'' + \alpha^2(\mathrm{i}\varphi' - 2f)] + -\frac{2}{3}\frac{\alpha^2}{R_L}\,(-f + \mathrm{i}\varphi') \tag{2}$$

$$\alpha^2\rho\,[\mathrm{i}(u-c)\varphi] = -\frac{\pi'}{\sigma M_1^2} + \frac{\alpha}{R_L}\,(2\varphi'' + \mathrm{i}f' - \alpha^2\varphi) + -\frac{2}{3}\frac{\alpha}{R_L}\,(\varphi'' + \mathrm{i}f') \tag{3}$$

$$\alpha\rho\,[\mathrm{i}(u-c)\theta + T'\varphi] = -\alpha\rho T(\sigma-1)(\varphi' + \mathrm{i}f) + \frac{\sigma}{R_L}\,(\theta'' - \alpha^2\theta) + \frac{\sigma(\sigma-1)M_1^2}{R_L}\,[2u'(f' + \mathrm{i}\alpha^2\varphi)] \tag{4}$$

$$\frac{\pi}{p} = \frac{\gamma}{\rho} + \frac{\theta}{T} \tag{5}$$

Note that the system of equations (1)–(5) has an essential singularity at $R_L \to \infty$. In principle, the solutions could be constructed as convergent-series expansions around one or more regular points, and these series could be properly joined to exponentially-decaying functions as $y \to \infty$. Since $(1/R_L)$ multiplies the highest-order derivatives, one might consider asymptotic expansions valid for $R_L \gg 1$.

(2) A SINGULAR-PERTURBATION PROBLEM

One notices in equations (1)–(5) that a small parameter multiplies one of the highest derivatives in the differential equations. Then a regular-perturbation scheme misses that derivative in the first approximation so that the order of the differential equations is reduced (this is consistent with the fact that the equations (1)–(5) have an essential singularity at $R_L \to \infty$). This approximation, under circumstances (to be pointed out in section (5)), breaks down in a thin layer centred at the point $u = c$ where the effect of viscosity persists no matter how large the Reynolds number be. The problem governed by equations (1)–(5) is clearly one of singular-perturbation type. In order to treat this problem, we use the method of matched asymptotic expansions, which allows for a systematic inclusion of the higher-order effects in successive approximations. In particular, the method of matched asymptotic expansions yields simplified equations of the same order of accuracy as the complete equations (1)–(5) for the inner viscous layered centred at $u = c$, but with fewer terms.

Assume the outer expansions as asymptotic representations

$$\begin{aligned}
\tilde{f} &\sim \Delta_1(R_L)\tilde{f}_1(y) + o(\Delta_1)\\
\tilde{\varphi} &\sim \Delta_1(R_L)\tilde{\varphi}_1(y) + o(\Delta_1)\\
\tilde{\pi} &\sim \Delta_1(R_L)\tilde{\pi}_1(y) + o(\Delta_1)\\
\tilde{\gamma} &\sim \Delta_1(R_L)\tilde{\gamma}_1(y) + o(\Delta_1)\\
\tilde{\theta} &\sim \Delta_1(R_L)\tilde{\theta}_1(y) + o(\Delta_1)
\end{aligned} \tag{6}$$

where $\Delta_1(R_L)$ must take on a finite value, the exact amount of which is evaluated later. Thus equations (1)–(5), for the first approximation to the outer solution, become

$$\mathrm{i}(u-c)\tilde{\gamma}_1 + \rho(\tilde{\varphi}_1' + \mathrm{i}\tilde{f}_1) + \rho'\tilde{\varphi}_1 = 0 \tag{7}$$

$$\rho[\mathrm{i}(u-c)\tilde{f}_1 + u'\tilde{\varphi}_1] = -\frac{\mathrm{i}\tilde{\pi}_1}{\sigma M_1^2} \tag{8}$$

$$\alpha^2\rho[\mathrm{i}(u-c)\tilde{\varphi}_1] = -\frac{\tilde{\pi}_1'}{\sigma M_1^2} \tag{9}$$

$$\rho[\mathrm{i}(u-c)\tilde{\theta}_1 + T'\tilde{\varphi}_1] = -\rho T(\sigma-1)(\tilde{\varphi}_1' + \mathrm{i}\tilde{f}_1) \tag{10}$$

$$\frac{\tilde{\pi}_1}{p} = \frac{\tilde{\gamma}_1}{\rho} + \frac{\tilde{\theta}_1}{T} \tag{11}$$

The outer system obtained above does not, under certain circumstances, represent the behaviour of the disturbances in the viscous-dominated regions centred at $u = c$ accurately. An inner solution then takes care of this viscous layer. One first needs to magnify the y-coordinate in this inner region so that the size of the inner region then remains independent of $(1/R_L)$ as $(1/R_L) \to 0$, and the width of the inner region becomes of order unity. This artifice leads to the retention of the highest derivatives in equations (1)–(5) as $(1/R_L) \to 0$, which are essential to represent the rapidly-varying behaviour of the disturbances in this viscous layer. Thus let

$$Y = \frac{y - y_c}{\Delta_1(R_L)}, \qquad \text{where } u(y_c) = c \tag{12}$$

Assume an inner expansion

$$\begin{aligned}
\hat{f} &\sim \Delta_1(R_L)\hat{f}_1(Y) + o(\Delta_1)\\
\hat{\varphi} &\sim \hat{\eta}_1(R_L)\hat{\varphi}_1(Y) + o(\hat{\eta}_1)\\
\hat{\pi} &\sim \Delta_1(R_L)\hat{\pi}_1(Y) + o(\Delta_1)\\
\hat{\gamma} &\sim \Delta_1(R_L)\hat{\gamma}_1(Y) + o(\Delta_1)\\
\hat{\theta} &\sim \Delta_1(R_L)\hat{\theta}_1(Y) + o(\Delta_1)
\end{aligned} \tag{13}$$

In order to deduce the symptotic matching conditions, between the first approximation to the outer and the inner solutions, write the outer expansions (6) near $y = y_c$ as

$$\begin{aligned}
\tilde{f} &= \Delta_1(R_L)\tilde{f}_1(y_c) + (y - y_c)\Delta_1(R_L)\tilde{f}_{1y}(y_c) + o(\Delta_1)\\
\tilde{\varphi} &= \Delta_1(R_L)\tilde{\varphi}_1(y_c) + (y - y_c)\Delta_1(R_L)\tilde{\varphi}_{1y}(y_c) + o(\Delta_1)\\
\tilde{\pi} &= \Delta_1(R_L)\tilde{\pi}_1(y_c) + (y - y_c)\Delta_1(R_L)\tilde{\pi}_{1y}(y_c) + o(\Delta_1)\\
\tilde{\gamma} &= \Delta_1(R_L)\tilde{\gamma}_1(y_c) + (y - y_c)\Delta_1(R_L)\tilde{\gamma}_{1y}(y_c) + o(\Delta_1)\\
\tilde{\theta} &= \Delta_1(R_L)\tilde{\theta}_1(y_c) + (y - y_c)\Delta_1(R_L)\tilde{\theta}_{1y}(y_c) + o(\Delta_1)
\end{aligned}$$

or in terms of the inner coordinates

$$\begin{aligned}
\tilde{f} &= \Delta_1(R_L)\tilde{f}_1(y_c) + \Delta_1^2(R_L)Y\tilde{f}_{1y}(y_c) + o(\Delta_1)\\
\tilde{\varphi} &= \Delta_1(R_L)\tilde{\varphi}_1(y_c) + \Delta_1^2(R_L)Y\tilde{\varphi}_{1y}(y_c) + o(\Delta_1)\\
\tilde{\pi} &= \Delta_1(R_L)\tilde{\pi}_1(y_c) + \Delta_1^2(R_L)Y\tilde{\pi}_{1y}(y_c) + o(\Delta_1)\\
\tilde{\gamma} &= \Delta_1(R_L)\tilde{\gamma}_1(y_c) + \Delta_1^2(R_L)Y\tilde{\gamma}_{1y}(y_c) + o(\Delta_1)\\
\tilde{\theta} &= \Delta_1(R_L)\tilde{\theta}_1(y_c) + \Delta_1^2(R_L)Y\tilde{\theta}_{1y}(y_c) + o(\Delta_1)
\end{aligned} \tag{14}$$

for a fixed y, as $(1/R_L) \to 0$, $Y \to \infty$. The above expressions therefore define the behaviour of the inner solution $\hat{f}$, $\hat{\varphi}$, $\hat{\pi}$, $\hat{\gamma}$, $\hat{\theta}$ as $Y \to \infty$. Then the conditions of asymptotic matching between the inner and the outer solutions (13), (14) give, as $Y \to \infty$,

$$\begin{aligned}\hat{f}_1(Y) &= \tilde{f}_1(y_c)\\ \hat{\pi}_1(Y) &= \tilde{\pi}_1(y_c)\\ \hat{\gamma}_1(Y) &= \tilde{\gamma}_1(y_c)\\ \hat{\theta}_1(Y) &= \tilde{\theta}_1(y_c)\end{aligned} \tag{15}$$

This presupposes that $\hat{f}_1(Y)$, $\hat{\pi}_1(Y)$, $\hat{\gamma}_1(Y)$, $\hat{\theta}_1(Y)$ approach a finite limit as $Y \to \infty$.

The deduction of the asymptotic matching conditions for φ_1 must be deferred until we are able to relate $\hat{\eta}_1(R_L0$ to $\Delta_1(R_L)$. In order to do it, substitute (13) into equations (1)–(5), so that there follows

$$\mathrm{i}(u-c)\Delta_1\hat{\gamma}_1 + \rho\left(\frac{\hat{\eta}_1\hat{\varphi}_1'}{\Delta_1} + \mathrm{i}\Delta_1\hat{f}_1\right) + \rho'\hat{\eta}_1\hat{\varphi}_1 = 0 \tag{16}$$

$$\begin{aligned}\alpha\rho[\mathrm{i}(u-u)\Delta_1\hat{f}_1 + u'\hat{\eta}_1\hat{\varphi}_1] &= -\frac{\mathrm{i}\alpha\Delta_1\hat{\pi}_1}{\sigma M_1^2} + \\ &+ \frac{1}{R_L}\left[\frac{\hat{f}_1''}{\Delta_1} + \alpha^2\left(\frac{\mathrm{i}\hat{\eta}_1\hat{\varphi}_1'}{\Delta_1} - 2\Delta_1\hat{f}_1\right)\right] + \\ &- \frac{2}{3}\frac{\alpha^2}{R_L}\left(-\Delta_1\hat{f}_1 + \frac{\mathrm{i}\hat{\eta}_1\hat{\varphi}_1'}{\Delta_1}\right)\end{aligned} \tag{17}$$

$$\begin{aligned}\alpha^2\rho[\mathrm{i}(u-c)\hat{\eta}_1\hat{\varphi}_1] &= -\frac{\hat{\pi}_1'}{\sigma M_1^2} + \\ &+ \frac{\alpha}{R_L}\left(2\frac{\hat{\eta}_1\hat{\varphi}_1''}{\Delta_1^2} + \mathrm{i}\hat{f}_1' - \alpha^2\hat{\eta}_1\hat{\varphi}_1\right) + \\ &- \frac{2}{3}\frac{\alpha}{R_L}\left(\frac{\hat{\eta}_1\hat{\varphi}_1''}{\Delta_1^2} + \mathrm{i}\hat{f}_1'\right)\end{aligned} \tag{18}$$

$$\begin{aligned}\alpha\rho[\mathrm{i}(u-c)\Delta_1\hat{\theta}_1 + T'\hat{\eta}_1\hat{\varphi}_1] &= -\alpha\rho T(\sigma-1)\left(\frac{\hat{\eta}_1\hat{\varphi}_1}{\Delta_1} + \mathrm{i}\Delta_1\hat{f}_1\right) + \\ &+ \frac{\alpha}{R_L}\left(\frac{\hat{\theta}_1''}{\Delta_1} - \alpha^2\Delta_1\hat{\theta}_1\right) + \\ &+ \frac{\sigma(\sigma-1)M_1^2}{R_L}[2u'(\hat{f}_1' + \mathrm{i}\alpha^2\hat{\eta}_1\hat{\varphi}_1)]\end{aligned} \tag{19}$$

$$\frac{\Delta_1\hat{\pi}_1}{p} = \frac{\Delta_1\hat{\gamma}_1}{\rho} + \frac{\Delta_1\hat{\theta}_1}{T} \tag{20}$$

From equation (16) it is clear that an appropriate choice (which is consistent with what Van Dyke (1975) calls 'principle of least degeneracy') is

$$\frac{\hat{\eta}_1}{\Delta_1} \sim \Delta_1$$

or

$$\hat{\eta}_1 \sim \Delta_1^2 \tag{21}$$

In order that the highest derivatives in equations (17) and (19) are retained, one requires as $R_L \to \infty$,

$$\Delta_1^2 R_L \sim 1$$

or

$$\Delta_1 \sim \frac{1}{\sqrt{R_L}} \tag{22}$$

The equations governing the first approximation to the inner solution then become

$$\mathrm{i}(u-c)\hat{\gamma}_1 + \rho(\hat{\varphi}_1' + \mathrm{i}\hat{f}_1) = 0 \tag{23}$$

$$\alpha\rho[\mathrm{i}(u-c)\hat{f}_1] = -\frac{\mathrm{i}\alpha\hat{\pi}_1}{\sigma M_1^2} + \hat{f}_1'' \tag{24}$$

$$\hat{\pi}_1' = 0 \tag{25}$$

$$\alpha\rho[\mathrm{i}(u-c)\hat{\theta}_1] = -\alpha\rho T(\sigma-1)(\hat{\varphi}_1' + \mathrm{i}\hat{f}_1) + \sigma\hat{\theta}_1'' \tag{26}$$

$$\frac{\hat{\pi}_1}{p} = \frac{\hat{\gamma}_1}{\rho} + \frac{\hat{\theta}_1}{T} \tag{27}$$

One is now ready to find the asymptotic matching conditions for φ_1. From comparing

$$\tilde{\varphi} = \Delta_1(R_L)\tilde{\varphi}_1(y_c) + \Delta_1^2(R_L)Y\tilde{\varphi}_{1y}(y_c) + o(\Delta_1)$$

and

$$\hat{\varphi} = \Delta_1^2(R_L)\hat{\varphi}_1(Y) + o(\Delta_1^2)$$

one finds in the limit $Y \to \infty$,

$$\tilde{\varphi}_1(y_c) = 0$$

$$\hat{\varphi}_1(Y) = Y\tilde{\varphi}_{1y}(y_c)$$

or

$$\tilde{\varphi}_1(y_c) = 0$$

$$\tilde{\varphi}_{1y}(y_c) = \tilde{\varphi}_{1Y}(Y) \tag{28}$$

Since ρ', T' no longer appear in euqations (23)–(26) the inner viscous layer centred at $u = c$ in effect has an incompressible flow.

Locally this situation has some of the characteristics of the oscillating-plate problem. Since the parameter $(\alpha R_{\mathrm{L}})^{-1/2}$ measures the relative diffusion distance for vorticity during one period, this inner boundary layer is thin when $\alpha R_{\mathrm{L}} \gg 1$. Thus one is led to adopt Prandtl's (1935) division of the disturbances into slowly-varying functions that are largely inviscid across the entire flow, and rapidly-varying functions near $u = c$ that are viscous.

It may be noted that equations (23)–(27) are different from a simplified set of equations based on some order-of-magnitude analysis proposed by Dunn and Lin (1955). The latter equations were used by Ng (1976) to give an asymptotic theory.

(3) TURNING-POINT PROBLEM

The first-order longitudinal velocity disturbances in the inner viscous layer centred at $u = c$ are governed by (recall equation (24))

$$\alpha\rho\,[\mathrm{i}(u-c)\hat{f}] = -\frac{\mathrm{i}\alpha\hat{\pi}}{\sigma M_1^2} + \frac{\hat{f}''}{R_{\mathrm{L}}} \qquad (29)$$

Differentiate equation (29) with respect to y^*; recall that in the inner viscous layer:

(i) the mean flow is effectively incompressible,
(ii) $\hat{\pi}' = 0$, and
(iii) the gradients of disturbances are much larger compared with absolute values of the disturbances – a feature verified in the Appendix to this section,

then one finds

$$\hat{f}''' + \lambda_1^2(y_c^* - y^*)\hat{f}' = 0 \qquad (30)$$

where

(a) in the neighbourhood of $u = c$, one writes

$$u = c + u_c' \cdot (y^* - y_c^*) + \ldots$$

(b) $|\lambda_1^2| = |\mathrm{i}\rho_c u_c' \alpha R_{\mathrm{L}}| \gg 1$.

Thus,

$$\left\{\left[\frac{\mathrm{d}^2}{\mathrm{d}y^{*2}} + \lambda_1^2(y_c^* - y^*)\right]\frac{\mathrm{d}}{\mathrm{d}y^*}\right\}\hat{f} = 0 \qquad (31)$$

One solution of the case is

$$\hat{f} \equiv 1 \tag{32}$$

which agrees with that in equation (110) (see next section). In order to find the other two solutions for $\hat{f}$, put

$$\hat{X} = \hat{f}', \quad y = \frac{y^*}{y_c^*}, \quad \lambda^2 = \lambda_1^2 y_c^{*3} \gg 1 \tag{33}$$

so that equation (31) becomes

$$\hat{X}'' + \lambda^2(1-y)\hat{X} = 0 \tag{34}$$

As we shall see in the following, the study of equation (34) reveals a new remarkable role of the critical point $u = c$, which now corresponds to a turning point of the mathematical problem.

We shall develop asymptotic expansions to represent the solutions of equation (34) by Horn's (1899) method. Such expansions are valid everywhere except at the turning point. We shall then find uniformly-valid expansions by using the method of matched asymptotic expansions.

(i) Horn's method

Write equation (34) as

$$\frac{d^2\hat{X}}{dy^2} + q(y;\lambda)\hat{X} = 0 \tag{35}$$

where

$$q(y;\lambda) = \lambda^2(1-y) \tag{36}$$

If,

$$q(y;\lambda) = \lambda^2 \sum_{n=0}^{\infty} \lambda^{-n} q_n(y) \quad \text{as } \lambda \to \infty \tag{37}$$

where $q_0 = (1-y) \neq 0$ in the interval of interest, the asymptotic solution can then be sought in the form

$$\hat{X}(y;\lambda) = e^{\lambda \sum_{n=0}^{\infty} \lambda^{-n} g_n(y)} \tag{38}$$

Substituting equations (37) and (38) into equation (35), and equating the coefficients of equal powers of λ, one finds

$$g_0'^2 + (1-y) = 0 \tag{39}$$

$$2g_0'g_1' + g_0'' = 0$$

Thus,

$$g_0 = \pm \mathrm{i} \int \sqrt{1-y}\, \mathrm{d}y$$

$$g_1 = -\tfrac{1}{2} \ln g_0' \tag{40}$$

Thence to first approximation,

$$\hat{X}_1 = \frac{a_1 \cos\left[\lambda \int \sqrt{1-y}\, \mathrm{d}y\right] + b_1 \sin\left[\lambda \int \sqrt{1-y}\, \mathrm{d}y\right]}{(1-y)^{1/4}} \quad \text{if } y < 1 \tag{41}$$

$$\hat{X}_2 = \frac{a_2\, \mathrm{e}^{\lambda \int \sqrt{y-1}\, \mathrm{d}y} + b_2\, \mathrm{e}^{-\lambda \int \sqrt{y-1}\, \mathrm{d}y}}{(y-1)^{1/4}} \quad \text{if } y > 1 \tag{42}$$

Note that these approximations are not valid near a point where $1 - y = 0$. Note that $\hat{X}(y)$ is oscillatory on one side of a zero of $(1 - y)$ whereas it is monotonic on the other side – the appropriate solution to equation (34) is given by (41) for $y < 1$, and by (42) for $y > 1$ – a feature which is physically compatible for the disturbances. Therefore such a zero of $(1 - y)$ is a transition point or a turning point. We shall now construct solutions that are valid in the neighbourhood of the turning point.

(ii) Method of matched asymptotic expansions

Note that the approximate solution to equation (34) is given by (41) for $y < 1$, and by (42) for $y > 1$; these are the outer asymptotic expansions which break down near the turning point $y = 1$. One then expects that a uniformly-valid expansion for all y must be expressed in terms of nonelementary functions which have the same qualitative features as the solutions of equation (34). In order to determine the magnitude of the region of nonuniformity of the straightforward expansions, put

$$\xi = \lambda^{\nu} (1-y), \quad \nu > 0 \tag{43}$$

so that equation (34) becomes

$$\frac{\mathrm{d}^2 \hat{X}}{\mathrm{d}\xi^2} + (\lambda^{2-3\nu} \xi)\, \hat{X} = 0 \tag{44}$$

It is obvious that according to the 'principle of least degeneracy' (Van Dyke, 1975) one requires, as $\lambda \to \infty$,

$$\nu = \tfrac{2}{3} \tag{45}$$

Note that this limit exhibits an exponential behaviour for $\xi < 0$ (i.e. $y > 1$, and an oscillatory behaviour for $\xi > 0$ (i.e. $y < 1$). Then one may use this limit as a 'crossing substitution' (Lin, 1945) for connecting the two solutions (41) and (42) as the turning point is crossed.

If one assumes an inner solution of the asymptotic form

$$\hat{X}^{\mathrm{i}}(\xi;\lambda) = \mu_0(\lambda)\,\hat{X}_0^{\mathrm{i}}(\xi) + o(\mu_0) \tag{46}$$

the zeroth-order inner solution is governed by

$$\frac{\mathrm{d}^2 \hat{X}_0^{\mathrm{i}}}{\mathrm{d}\xi^2} + \xi\,\hat{X}_0^{\mathrm{i}} = 0 \tag{47}$$

of which a general solution is

$$\hat{X}_0^{\mathrm{i}}(\xi) = \sqrt{\xi}\left[c_1 J_{1/3}\left(\tfrac{2}{3}\xi^{3/2}\right) + c_2 J_{-1/3}\left(\tfrac{2}{3}\xi^{3/2}\right)\right] \tag{48}$$

where $J_n(x)$ is Bessel's function of order n. Equation (48) becomes as $y \to 1$, since $J_\nu(t) \to t^\nu + o(t^\nu)$ as $t \to 0$,

$$\hat{\chi}_0^{\mathrm{i}} \approx C_1 + \mathrm{D}_1 y \tag{49}$$

so that this solution is uniformly valid (i.e. $\hat{X}_0^{\mathrm{i}}$ is bounded and well behaved and has no branches for all y, including the neighbourhood of the turning point $y = 1$).

In order to match the inner solution (48) to the outer solution (41) corresponding to $y < 1$, express (41) in terms of $\xi = \lambda^{2/3}(1 - y)$ and note that

$$\lambda \int_y^1 \sqrt{1-\tau}\,\mathrm{d}\tau = \frac{2}{3}\xi^{3/2} + O(\lambda^{-2/3})$$

so that as $\lambda \to \infty$,

$$\hat{X}^{\mathrm{o}} = \frac{\lambda^{1/6}}{\xi^{1/4}}\left[a_1 \cos\left(\frac{2}{3}\xi^{3/2}\right) + b_1 \sin\left(\frac{2}{3}\xi^{3/2}\right)\right] + \ldots \tag{50}$$

And for $\xi > 0$, $\xi \gg 1$, (48) gives

$$\hat{X}^{\mathrm{i}} = \mu_0(\lambda)\,\frac{\xi^{-1/4}}{\sqrt{\pi}}\left[a_3 \sin\left(\frac{2}{3}\xi^{3/2} + \frac{\pi}{4}\right) + {}\right.$$

$$\left. + b_3 \cos\left(\frac{2}{3}\xi^{3/2} + \frac{\pi}{4}\right)\right] + \ldots \tag{51}$$

The asymptotic matching between (50) and (51) gives

$$\mu_0(\lambda) = \lambda^{1/6}$$

$$a_1 = \frac{1}{\sqrt{\pi}}\left[a_3 \sin\frac{\pi}{4} + b_3 \cos\frac{\pi}{4}\right]$$

$$b_1 = \frac{1}{\sqrt{\pi}}\left[a_3 \cos\frac{\pi}{4} - b_3 \sin\frac{\pi}{4}\right] \tag{52}$$

Therefore, (41) becomes

$$\hat{X}^{o}_{01} = \frac{1}{\sqrt{\pi}(1-y)^{1/4}} \left\{ b_3 \cos\left[\lambda \int_y^1 \sqrt{1-\tau}\, d\tau + \frac{\pi}{4}\right] + \right.$$

$$\left. + a_3 \sin\left[\lambda \int_y^1 \sqrt{1-\tau}\, d\tau + \frac{\pi}{4}\right] \right\} \tag{53}$$

In order to match (48) to (42) corresponding to $y > 1$, express (42) in terms of $-\xi = \lambda^{2/3}(y-1)$ and note

$$\lambda \int_1^y \sqrt{\tau - 1}\, d\tau = \frac{2}{3}(-\xi)^{3/2} + O(\lambda^{-2/3})$$

so that as $\lambda \to \infty$,

$$\hat{X}^{o} = \frac{\lambda^{1/6}}{(-\xi)^{1/4}} \left[a_2\, e^{\frac{2}{3}(-\xi)^{3/2}} + b_2\, e^{-\frac{2}{3}(-\xi)^{3/2}} \right] + \ldots \tag{54}$$

And for $\xi < 0$, $|\xi| \gg 1$, (48) gives

$$\hat{X}^{i} = \mu_0(\lambda) \frac{(-\xi)^{1/4}}{\sqrt{\pi}} \left[\frac{a^3}{2}\, e^{-\frac{2}{3}(-\xi)^{3/2}} + b_3\, e^{\frac{2}{3}(-\xi)^{3/2}} \right] \tag{55}$$

The asymptotic matching between (54) and (55) gives

$$\mu_0(\lambda) = \lambda^{1/6}$$

$$a_2 = b_3/\sqrt{\pi}$$

$$b_2 = a_3/\sqrt{\pi} \tag{56}$$

Therefore, (42) becomes

$$\hat{X}^{o}_{02} = \frac{1}{\sqrt{\pi}(y-1)^{1/4}} \left[b_3\, e^{\lambda \int_1^y \sqrt{\tau-1}\, d\tau} + \right.$$

$$\left. + \frac{a_3}{2}\, e^{-\lambda \int_1^y \sqrt{\tau-1}\, d\tau} \right] \tag{57}$$

Thus an approximate solution to equation (34) with a turning point is given by the three separate expansions:

(i) near the turning point $y = 1$;
(ii) for $y < 1$;
(iii) for $y > 1$.

Note the connection between the constants a_1, b_1 and a_2, b_2 provided by the asymptotic matching conditions – a result first found by Rayleigh (1912) while investigating the total reflection of sound waves from a transition layer.

(iii) Stokes' Phenomenon

For the inviscid problem, the critical point $y = y_c$, where $u = c$, in the complex y-plane is a regular singularity. For the viscous problem, the critical point corresponds to the turning point of the mathematical problem. We shall see in the following that this latter feature leads the critical point to become a seat of Stokes' phenomenon – a result which has important physical implications, as pointed out later. In order to connect the two approximations on either side of $y = 1$ (so that they correspond to the same exact solution) we used previously an approximate solution which is valid along the entire length of some path connecting the regions of y on either side of $y = 1$, where the above approximations are valid on the real axis through $y = 1$.

Alternatively, one may avoid the real axis near $y = 1$, and use a path encircling $y = 1$ in the complex y-plane, along which the above two straightforward approximations themselves remain valid – a procedure due to Kemble (1958).

Let us introduce

$$\psi_{1,2} = \frac{e^{\pm i\lambda \int_1^y \sqrt{1-y}\, dy}}{(1-y)^{1/4}} \tag{58}$$

where the right-hand sides are to be considered as functions of the complex variable y, and one must introduce a 'cut' along the negative real axis of the complex y-plane in order to uniquely define $\sqrt{1-y}$.

Note,

$$\psi_{1,2}'' + \left[\lambda^2(1-y) - \frac{5}{16}\frac{1}{(1-y)^2}\right]\psi_{1,2} = 0 \tag{59}$$

Write

$$\hat{X}(y) = a_1(y)\,\psi_1(y) + a_2(y)\,\psi_2(y) \tag{60}$$

where $a_1(y), a_2(y)$ are the so-called osculating parameters so that

$$\hat{X}_1'(y) = a_1(y)\,\psi_1'(y) + a_2(y)\,\psi_2'(y) \tag{61}$$

From equations (60) and (61), one finds

$$a_1 = \frac{\hat{X}\psi_2' - \hat{X}'\psi_2}{\psi_1\psi_2' - \psi_1'\psi_2}, \quad a_2 = -\frac{\hat{X}\psi_1' - \hat{X}'\psi_1}{\psi_1\psi_2' - \psi_1'\psi_2}$$

and using (58),

$$\psi_1\psi_2' - \psi_1'\psi_2 = -2i$$

so that

$$a_1 = \frac{i}{2}(\hat{X}\psi_2' - \hat{X}'\psi_2), \quad a_2 = -\frac{i}{2}(\hat{X}\psi_1' - \hat{X}'\psi_1) \tag{62}$$

From which upon using equations (59) and (34), one finds

$$\frac{da_{1,2}}{dy} = \pm \frac{i}{2}(\hat{X}\psi''_{2,1} - \hat{X}''\psi_{2,1})$$

or

$$\frac{da_{1,2}}{dy} = \pm \frac{5i}{32}\frac{1}{(1-y)^2}\hat{X}\psi_{2,1}$$

Using (58), (60), this becomes

$$\frac{da_{1,2}}{dy} = \pm \frac{5i}{32}\frac{1}{(1-y)^{5/2}}\left[a_{1,2} + a_{2,1}\, e^{\mp 2i\lambda \int_y^1 \sqrt{1-y}\, dy}\right] \tag{63}$$

Integrating equation (63) along a path Γ in the complex y-plane from y_1 to y_2,

$$\Delta a_{1,2} = \int_\Gamma \frac{da_{1,2}}{dy} dy$$

$$= \pm \frac{5i}{16}\int_\Gamma \frac{a_{1,2}}{(1-y)^{5/2}} dy \pm \frac{5i}{16}\int_\Gamma \frac{a_{2,1}}{(1-y)^{5/2}}\left(\frac{\psi_{2,1}}{\psi_{1,2}}\right) dy \tag{64}$$

If we consider only such paths Γ that

$$\int_\Gamma \left|\frac{1}{(1-y)^{5/2}}\right| |dy| \ll 1$$

then along Γ, the first term on the right-hand side of (64) does not contribute significantly to $\Delta a_{1,2}$. The second term can be significant only if

$$\left|\frac{\psi_{2,1}}{\psi_{1,2}}\right| \gg 1$$

Noting that

$$\int_y^1 \sqrt{1-y}\, dy = \frac{2}{3}(1-y)^{3/2}$$

or

$$\int_y^1 \sqrt{1-y}\, dy = \frac{2}{3}[(1-y_r)^2 + y_i^2]^{3/4}$$

$$\times \left[\cos\frac{3}{2}\left(\tan^{-1}\frac{-y_i}{1-y_r}\right) + i\sin\frac{3}{2}\left(\tan^{-1}\frac{-y_i}{1-y_r}\right)\right] \tag{65}$$

One sees that, as $y_i \to 0$, it is only in the neighbourhood of $y = 1$ that it is possible to have

$$\left| \frac{\psi_{2,1}}{\psi_{1,2}} \right| \gg 1$$

so that the coefficient of a very small function may vary remarkably – a typical feature of Stokes' phenomenon (another feature of which will be discussed in section (5)) at the turning point. Physically this feature is traceable, for instance, in the problem of the generation of surface waves in water by a wind blowing over it (see Chapter VII) to the fact that the energy transfer from the wind to the surface waves in water occurs primarily in the thin layer centred at the critical point $u = c$.

(iv) Appendix

In studying the turning-point problem associated with equation (29) it was stated as an assumption that the gradients of disturbances are much larger compared with the absolute values of the disturbances. We now proceed to verify that this indeed is the case. Differentiate equation (29) with respect to y^*, recall that in the inner layer

(i) the mean flow is effectively incompressible,
(ii) $\hat{\pi}' = 0$,

then one finds

$$\hat{f}''' + \lambda^2 \, [(1 - y)\hat{f}' - \hat{f}] = 0 \tag{66}$$

where

$$y - \frac{y^*}{y_c^*}, \qquad |\lambda^2| = |i\rho_c u_c' \alpha R_L y_c^{*3}| \gg 1 \tag{67}$$

Put

$$z = 1 - y \tag{68}$$

so that equation (66) becomes

$$\hat{f}''' + \lambda^2 z \hat{f}' + \lambda^2 \hat{f} = 0 \tag{69}$$

which we study here using a theory developed by Langer (1955).

Equation (69) has associated with it a characteristic equation

$$\chi^3 + z\chi = 0 \tag{70}$$

The structure of the solutions of equation (69) in a given region of the complex z-plane depends in an essential way upon the configuration of the roots of equation (70) in that region. If the roots are distinct throughout the region, or if they are distinct except for coincidences that prevail identically over the region, the forms of the solutions, as they depend asymptotically upon λ, are determinable by the well-known theories. Exception occurs if the region of z is one within which a coincidence of roots of equation (70) occurs at an isolated point – called the turning point of equation (69). In the neighbourhood of such a point, the z-region is divided into parts in which the solutions have distinctive forms, which are interrelated as we saw in section (3).

We now develop formal solutions of equation (69) in powers of $(1/\lambda)$. The roots of equation (70) are

$$\chi_0 = 0, \quad \chi_1 = \mathrm{i}z^{1/2}, \quad \chi_2 = -\mathrm{i}z^{1/2} \tag{71}$$

Correspondingly, following Langer (1955), put

$$\hat{f} = k\,\mathrm{e}^{\lambda\int\chi\,\mathrm{d}z}\sum_{n=0}^{\infty}\frac{\theta_n(z)}{\lambda^n} \tag{72}$$

so that one finds from equation (69),

$$\sum_{n=0}^{\infty}\frac{\lambda^2(3\chi^2+z)\theta_n'+\lambda^2(3\chi\chi'+1)\theta_n+ +\lambda(3\chi\theta_n''+3\chi'\theta_n'+\chi''\theta_n)+\theta_n''}{\lambda^n}=0$$

or

$$\sum_{n=0}^{\infty}\frac{(3\chi^2+z)\theta_n'+(3\chi\chi'+1)\theta_n+ +(3\chi\theta_{n-1}''+3\chi'\theta_{n-1}'+\chi''\theta_{n-1}+\theta_{n-2}'')}{\lambda^n}=0 \tag{73}$$

it being understood that $\theta_m = 0$ for $m < 0$. Thus,

$$(3\chi^2+z)\theta_n'+(3\chi\chi'+1)\theta_n+ = -(3\chi\theta_{n-1}''+3\chi'\theta_{n-1}'+\chi''\theta_{n-1}+\theta_{n-2})\; n=0,1,2,\ldots \tag{74}$$

Corresponding to

$$\chi = \chi_0 = 0$$

equation (74) becomes

$$z\theta_n'+\theta_n = -\theta_{n-2}''; \quad n=0,1,2,\ldots \tag{75}$$

For which,

$$\theta_{2n+1} \equiv 0$$
$$\theta_{2n} = K_n z^{-1-3n} \tag{76}$$

with

$$K_0 = 1$$
$$K_n = -\frac{3^2 n\,(n-\frac{1}{3})(n-\frac{2}{3})}{n}K_{n-1} \tag{77}$$

so that

$$\hat{f}_0 = kz^{-1} \sum_{n=0}^{\infty} \frac{K_n}{\lambda^{2n} z^{3n}} \tag{78}$$

Corresponding to

$$\chi = \chi_1, \quad \chi_2 = \pm\, \mathrm{i} z^{1/2}$$

equation (74) becomes

$$2z\theta_n' - \frac{1}{2}\theta_n = \frac{\pm \mathrm{i}}{z^{1/2}} \left[3z\,\theta''_{n-1} + \frac{3}{2}\theta'_{n-1} - \frac{\theta_{n-1}}{4z} \right] + \\ + \theta''_{n-2}; \quad n = 0, 1, 2, \ldots \tag{79}$$

for which

$$0_n = (\pm 1)^n \, \nu_n \, z^{\frac{3}{2}(-n-\frac{1}{6})} \tag{80}$$

with

$$\nu_0 = 1$$

$$\nu_1 = -\frac{\mathrm{i}}{?4}$$

$$\nu_n = \frac{\mathrm{i}[4 - 3(6n-3)(6n-5)]}{48} \nu_{n-1} + \\ + \frac{(6n-3)(6n-7)(6n-11)}{192n} \nu_{n-2} \quad n = 2, 3, \ldots \tag{81}$$

so that

$$\hat{f}_{1,2} = kz^{-1/4}\, \mathrm{e}^{\pm \frac{2}{3}\mathrm{i}\lambda z^{3/2}} \sum_{n=0}^{\infty} \frac{\nu_n}{(\pm\lambda z^{3/2})^n} \tag{82}$$

Put

$$\xi = \frac{1}{3}\mathrm{i}\lambda z^{3/2} \tag{83}$$

so that equations (78) and (82) become, respectively,

$$\hat{f}_0 = \xi^{-2/3} \sum_{n=0}^{\infty} \frac{K_n}{(3\mathrm{i}\xi)^{2n}}$$

$$\hat{f}_{1,2} = \xi^{-1/6}\, \mathrm{e}^{\pm 2\xi} \sum_{n=0}^{\infty} \frac{\nu_n}{(\mp 3\mathrm{i}\xi)^{2n}} \tag{84}$$

which are formal solutions of equation (69). The forms of the solutions of equation (69) with respect to ξ, corresponding to

$$(q-1)\pi \leqslant \arg(\xi) \leqslant q\,\pi, \quad q \text{ an integer} \tag{85}$$

are

$$\hat{f}_{q,0}(z) \sim \xi^{-2/3} \sum_{n=0}^{\infty} \frac{K_n}{(3\mathrm{i}\xi)^{2n}}$$

$$\hat{f}_{q,1}(z) \sim \xi^{-1/6}\, \mathrm{e}^{2(-1)^q \xi} \sum_{n=0}^{\infty} \frac{\nu_n}{[-3\mathrm{i}(-1)^q \xi]^{2n}}$$

$$\hat{f}_{q,2}(z) \sim \xi^{-1/6}\, \mathrm{e}^{-2(-1)^q \xi} \sum_{n=0}^{\infty} \frac{\nu_n}{[3\mathrm{i}(-1)^q \xi]^{2n}} \tag{86}$$

The agreement of the solutions $\hat{f}_{q,1}(z)$, $\hat{f}_{q,2}(z)$ in (86) with those developed previously verifies the validity of the assumption made at the beginning of this section that the gradients of disturbances are much larger compared with the absolute values of the disturbances.

Indeed, if one writes equation (69) as

$$\frac{\mathrm{d}}{\mathrm{d}z}[\hat{f}'' + \lambda^2 z \hat{f}] = 0 \tag{87}$$

which admits as solutions those of the equation

$$\hat{f}'' + \lambda^2 z \hat{f} = 0 \tag{88}$$

and these are known to be given in terms of thc Hankel functions (or the Bessel functions) of order 1/3:

$$\hat{f}_{q,1}(z) = \sqrt{\pi}\, \mathrm{e}^{-(q/2+1/6)\pi \mathrm{i}}\, \xi^{1/3}\, H^{(2)}_{1/3}(2\mathrm{i}\xi\, \mathrm{e}^{-q\pi \mathrm{i}})$$

$$\hat{f}_{q,2}(z) = \sqrt{\pi}\, \mathrm{e}^{-(q/2-2/3)\pi \mathrm{i}}\, \xi^{1/3}\, H^{(1)}_{1/3}(2\mathrm{i}\xi\, \mathrm{e}^{-q\pi \mathrm{i}}) \tag{89}$$

which, as we shall see in the next section, agrees with the asymptotic form of the convergent-series solution developed for the full equations (1)–(5).

(4) CONVENTIONAL VISCOUS SOLUTIONS

The conventional viscous solutions can be constructed by several interrelated methods introduced by Heisenberg (1924) for the case of an incompressible fluid:

(i) the WKB method leads to solutions of the exponential type, which are valid in certain regions of the complex y-plane excluding the immediate neighbourhood of the critical point;

(ii) in the second method, the solutions are obtained in terms of modified Hankel's functions which are valid in a complex neighbourhood of the critical point; these solutions may also be used to determine the correct branch of the WKB solutions in the neutral case.

(i) Conventional-series solution

Note that the solutions of equations (1)–(5) are analytic functions of y, wherever $u(y)$, $\rho(y)$, $T(y)$ are analytic, and are entire functions of α, c, R_L. However, the quantity $1/R_L$ multiplies the highest derivatives in equations (1)–(5), so that as we have previously pointed out a straightforward expansion in terms of the parameter $1/R_L$ will not be uniformly valid. An alternative method is to choose a small parameter ϵ related to $(\alpha R_L)^{-1/3}$, and make a transformation

$$y - y_c = \epsilon\eta, \qquad \epsilon = (\alpha R_L)^{-1/3} \tag{90}$$

where

$$u(y_c) = c$$

Then equations (1)–(5) become

$$\alpha\rho_c[\mathrm{i}u_c'\epsilon\eta f + u_c'\varphi] = -\frac{\mathrm{i}\alpha\pi}{\sigma M_1^2} +$$

$$+\,\epsilon\alpha[f'' + \alpha^2(\mathrm{i}\epsilon\varphi' - 2\epsilon^2 f)] - \frac{2\alpha}{3}\epsilon(-\epsilon^2 f + \mathrm{i}\epsilon\varphi) \tag{91}$$

$$\alpha^2\rho_c[\mathrm{i}u_c'\epsilon\eta\varphi] = -\frac{\pi'}{\sigma M_1^2} + \epsilon\alpha^2(2\varphi'' - \mathrm{i}\epsilon f' - \alpha^2\epsilon^2\varphi) +$$

$$-\frac{2\alpha^2}{3}\epsilon(\varphi'' + \mathrm{i}\epsilon f') - \frac{2\alpha}{3}\epsilon(\epsilon\varphi' + \mathrm{i}\epsilon^2 f) \tag{92}$$

$$\alpha\rho_c[\mathrm{i}u_c'\epsilon\eta\theta + T_c'\varphi] = -\alpha\rho_c T_c(\sigma - 1)\left(\frac{\varphi'}{\epsilon} + \mathrm{i}f\right) +$$

$$+\,\epsilon\sigma(\theta'' - \alpha^2\epsilon^2\theta) +$$

$$+\,\sigma(\sigma - 1)\,\epsilon M_1^2\,[2u_c'(\epsilon f' + \mathrm{i}\alpha^2\epsilon^2\varphi)] \tag{93}$$

$$\frac{\varphi'}{\epsilon} + \mathrm{i}f - \frac{T_c'}{T_c}\varphi + \mathrm{i}u_c'\,\epsilon\eta\left(\frac{\pi}{\rho_c T_c} - \frac{\theta}{T_c}\right) = 0 \tag{94}$$

where

(i) the mean-flow quantities u, ρ, T have been represented in the neighbourhood of $y = y_c$ by Taylor's series:

$$u - c = u_c'\,\epsilon\eta + \frac{u_c''}{2!}(\epsilon\eta)^2 + \ldots \tag{95}$$

$$\rho = \rho_c + \rho_c' \cdot \epsilon\eta + \ldots$$

$$T = T_c + T_c' \cdot \epsilon\eta + \ldots$$

(ii) γ has been eliminated by using equation (5);
(iii) the primes now denote differentiation with respect to η.

The coefficients in equations (91)–(94) are therefore convergent power series in ϵ as long as the power series given in (95) are convergent. One then attempts to obtain a fundamental system of solutions as powers of series ϵ, of the form

$$
\begin{aligned}
f &= X_1^{(0)}(\eta) + \epsilon X_1^{(1)}(\eta) + \ldots \\
\varphi &= \epsilon X_2^{(0)}(\eta) + \epsilon^2 X_2^{(1)}(\eta) + \ldots \\
\pi &= X_3^{(0)}(\eta) + \epsilon X_3^{(1)}(\eta) + \ldots \\
\theta &= X_4^{(0)}(\eta) + \epsilon X_4^{(1)}(\eta) + \ldots
\end{aligned} \tag{96}
$$

Substitute (96) into equations (91)–(94), and comparing the coefficients of equal powers of ϵ, one obtains

$$\alpha\rho_c[\mathrm{i}u_c'\eta X_1^{(0)} + u_c' X_2^{(0)}] = \alpha X_1^{(0)''} \tag{97}$$

$$\alpha^2\rho_c[\mathrm{i}u_c'\eta X_2^{(0)}] = \alpha^2(2X_2^{(0)''} - \mathrm{i}X_1^{(0)'}) + \\ - \frac{2\alpha^2}{3}(X_2^{(0)''} + \mathrm{i}X_1^{(0)'}) \tag{98}$$

$$0 = -\frac{X_3^{(0)'}}{\sigma M_1^2} \tag{99}$$

$$\alpha\rho_c[\mathrm{i}u_c'\eta X_4^{(0)} + T_c' X_2^{(0)}] = X_4^{(0)''} \tag{100}$$

$$X_2^{(0)} + \mathrm{i}X_1^{(0)} = 0 \tag{101}$$

From equations (97)–(101), one obtains

$$X_1^{(0)'''} - \mathrm{i}\rho_c u_c'\eta X_1^{(0)'} = 0 \tag{102}$$

which can be transformed to Bessel's equation by putting

$$\zeta = \eta^{3/2}(\rho_c u_c')^{1/2} \tag{103}$$

so that

$$\left(\frac{\mathrm{d}^2}{\mathrm{d}\zeta^2} + \frac{1}{\zeta}\frac{\mathrm{d}}{\mathrm{d}\zeta} + \frac{4\mathrm{i}^3}{9} - \frac{1}{9\zeta^2}\right)\frac{\mathrm{d}X_1^{(0)}}{\mathrm{d}\zeta} = 0 \tag{104}$$

The solutions of this equation are

$$\frac{\mathrm{d}X_{11}^{(0)}}{\mathrm{d}\zeta} = H_{1/3}^{(1)}\left[\frac{2}{3}(\mathrm{i}^{3/2}\zeta)\right] \tag{105}$$

$$\frac{\mathrm{d}X_{12}^{(0)}}{\mathrm{d}\zeta} = H_{1/3}^{(2)}\left[\frac{2}{3}(\mathrm{i}^{3/2}\zeta)\right] \tag{106}$$

$$\frac{dX_{13}^{(0)}}{d\zeta} = 0 \tag{107}$$

or

$$X_{11}^{(0)} = \beta^{3/2} \int_{\infty}^{\zeta} H_{1/3}^{(1)} \left[\frac{2}{3}(i^{3/2}\zeta)\right] d\zeta \tag{108}$$

$$X_{12}^{(0)} = \beta^{3/2} \int_{\infty}^{\zeta} H_{1/3}^{(2)} \left[\frac{2}{3}(i^{3/2}\zeta)\right] d\zeta \tag{109}$$

$$X_{13}^{(0)} = 1 \tag{110}$$

where

$$\beta = \rho_c u_c'$$

Then,

$$X_{21}^{(0)} = \left\{-i\zeta^{2/3} \int_{\infty}^{\zeta} H_{1/3}^{(1)} \left[\frac{2}{3}(i^{3/2}\zeta)\right] d\zeta + \right.$$

$$\left. + i \int_{\infty}^{\zeta} \zeta^{2/3} H_{1/3}^{(1)} \left[\frac{2}{3}(i^{3/2}\zeta)\right] d\zeta \right\} \frac{1}{\sqrt{\rho_c u_c'}} \tag{111}$$

$$X_{22}^{(0)} = \left\{-i\zeta^{2/3} \int_{\infty}^{\zeta} H_{1/3}^{(2)} \left[\frac{2}{3}(i^{3/2}\zeta)\right] d\zeta + \right.$$

$$\left. + i \int_{\infty}^{\zeta} \zeta^{2/3} H_{1/3}^{(2)} \left[\frac{2}{3}(i^{3/2}\zeta)\right] d\zeta \right\} \frac{1}{\sqrt{\rho_c u_c'}} \tag{112}$$

$$X_{23}^{(0)} = -i \left(\frac{1}{\rho_c u_c'}\right)^{1/3} \zeta^{2/3} \tag{113}$$

Here $H_n^{(1)}(x)$, and $H_n^{(2)}(x)$ are Hankel's functions of the first kind and second kind respectively.

It may be noted that the choice of Hankel's functions as solutions for the disturbances is appropriate since the former have an asymptotic behaviour:

$$H_n^{(1)}(iz) \sim \sqrt{\frac{2}{\pi z}}\, e^{-z}, \quad \text{as } z \to \infty$$

$$H_n^{(2)}(iz) \sim \sqrt{\frac{2}{\pi z}}\, e^{z}, \quad \text{as } z \to \infty \tag{114}$$

so that one rejects the solution given in terms of $H_{1/3}^{(2)}\left[\frac{2}{3} i^{3/2}\zeta\right]$ because of the requirements of boundedness at infinity.

(ii) Asymptotic-series solution

In analogy with the incompressible case, we construct here two asymptotic solutions for the equations governing the inner viscous layer by expanding the solution in powers of $(\alpha R_L)^{-1}$ – a procedure which is reminiscent of the Wentzel–Brillouin–Kramers method. These asymptotic-series solutions reflect the effect of viscosity, however, for large Reynolds number. Put

$$\hat{f}_1(y) = \left[e^{\pm\hat{G}(y)\sqrt{\alpha R_L}}\right]\left[\hat{g}_0(y) + \frac{1}{\sqrt{\alpha R_L}}\hat{g}_1(y) + \ldots\right] \tag{115}$$

where $\hat{G}(y)$, $\hat{g}_0(y)$, $\hat{g}_1(y)$, . . . are obtained by formal substitution into equation (24), and comparing the coefficients. First, differentiate equation (24) with respect to y and use equation (25) to obtain

$$\alpha\rho[\mathrm{i}(u-c)]\hat{f}' + [\alpha\rho'\mathrm{i}(u-c) + \alpha\rho\,\mathrm{i}u']\hat{f} = -\hat{f}'''/R_L \tag{116}$$

Thus,

$$\alpha\rho\Bigg[\mathrm{i}(u-c)\Bigg\{e^{\pm\sqrt{\alpha R_L}\hat{G}(y)}\left[\hat{g}_0'(y) + \frac{1}{\sqrt{\alpha R_L}}\hat{g}_1'(y) + \ldots\right] +$$

$$+\sqrt{\alpha R_L}\,\hat{G}'(y)\,e^{\pm\sqrt{\alpha R_L}\hat{G}(y)}\left[\hat{g}_0(y) + \frac{1}{\sqrt{\alpha R_L}}\hat{g}_1(y) + \ldots\right]\Bigg\}\Bigg]$$

$$+[\alpha\rho'\mathrm{i}(u-c) + \alpha\rho'\mathrm{i}u']\Bigg[e^{\pm\sqrt{\alpha R_L}\hat{G}(y)}\Bigg\{\hat{g}_0(y) +$$

$$+\frac{1}{\sqrt{\alpha R_L}}\hat{g}_1(y) + \ldots\Bigg\}\Bigg]$$

$$= \frac{1}{R_L}\Bigg[e^{\pm\sqrt{\alpha R_L}\hat{G}(y)}\left\{\hat{g}_0'''(y) + \frac{1}{\sqrt{\alpha R_L}}\hat{g}_1'''(y) + \ldots\right\} +$$

$$+\left\{3\sqrt{\alpha R_L}\,\hat{G}'(y)\right\}\left\{e^{\pm\sqrt{\alpha R_L}\hat{G}(y)}\right\}\left\{\hat{g}_0''(y) + \frac{1}{\sqrt{\alpha R_L}}\hat{g}_1''(y) + \ldots\right\}$$

$$+\left\{3\sqrt{\alpha R_L}\,\hat{G}''(y) + 3\alpha R_L\,\hat{G}'^2(y)\right\}\left\{e^{\pm\sqrt{\alpha R_L}\hat{G}(y)}\right\}\times$$

$$\times\left\{\hat{g}_0'(y) + \frac{1}{\sqrt{\alpha R_L}}\hat{g}_1'(y) + \ldots\right\} +$$

$$+\left\{3\alpha R_L\,\hat{G}'(y)\,\hat{G}''(y) + (\alpha R_L)^{3/2}\,\hat{G}'^3(y) +\right.$$

$$\left.+\sqrt{\alpha R_L}\,\hat{G}'''(y)\right\}\left\{e^{\pm\sqrt{\alpha R_L}\hat{G}(y)}\right\}\times$$

$$\times\left\{\hat{g}_0(y)+\frac{1}{\sqrt{\alpha R_L}}\,\hat{g}_1(y)+\ldots\right\}\Bigg] \tag{117}$$

From which

$$\hat{G}(y)=\int_{y_c}^{y}\sqrt{\mathrm{i}\rho(u-c)}\,\mathrm{d}y \tag{118}$$

And $\hat{g}_0(y)$ is given by comparison of equation (115) with the asymptotic limit of the convergent-series solution (108) valid in the complex neighbourhood of the point $u = c$; thus

$$\hat{g}_0(y)=(u-c)^{-3/4} \tag{119}$$

The domain of validity and the proper branch of the multiple-valued asymptotic solutions (115) can be revealed by a careful examination of the manner in which the latter represent valid asymptotic approximations to the exact solutions of equations (1)–(5).

Note that the WKB solutions (IV.26) are not valid in a domain of the complex y-plane that contains the critical point $u = c$, nor do they give the boundary values of the solutions sufficiently accurately if $|y-y_c|\sim(\alpha R_L)^{-1/3}$. The point $u = c$ appears in the inviscid solutions as a logarithmic branch point. Here it appears as an algebraic point. Again the determination of the correct branch so as to achieve the analytical continuation of the functions across the point $u = c$ should be such that (115) yields valid asymptotic solutions of equations (1)–(5) all along the path. This is necessary *a fortiori* if the critical point y_c lies below or on the real axis joining y_1, y_2 where the boundary conditions are to be satisfied. One must therefore take a path from y_1 to y_2 in the complex y-plane such that the real part of $\hat{G}(y)$ increases monotonically. It turns out (see next section) that if one takes a path below the critical point y_c, one is then always in a region in the complex y-plane where the above asymptotic solutions are valid (see Fig. 1). In fact, if $c_i > 0$ (however let $c_i \ll 1$)

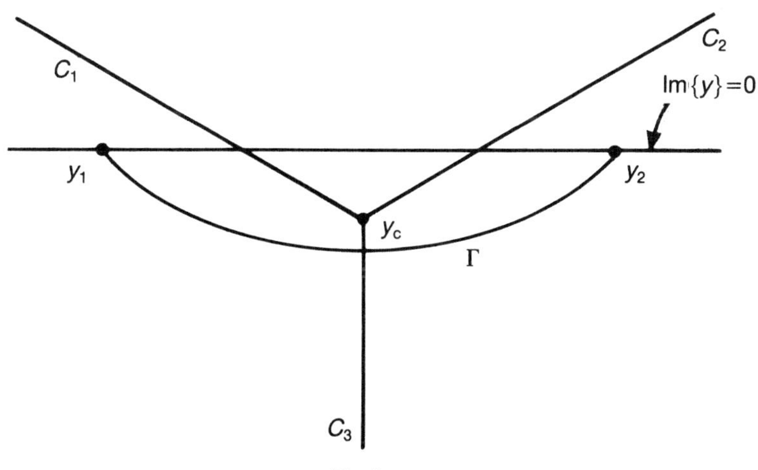

Fig. 1

and $u_c' > 0$, as we shall see in the next section, the point y_c lies above the real axis, and the asymptotic solutions (115) are valid all along the real axis of y. For $c_i = 0$, the point y_c lies on the real axis of y where the asymptotic solutions (115) cease to be valid. For $c_i < 0$, the point y_c is below the real axis, and one can use the solutions (115) only along the part of the real axis and their analytic continuation must be determined by a path circling below the critical point.

Finally if $|y - y_c| \ll 1$, but still large compared with $(\alpha R_L)^{-1/3}$, then the asymptotic solutions can be further simplified to

$$\hat{f}_1(y) \sim [u_c'(y - y_c)^{-3/4}] \times \times \left[e^{\pm\frac{2}{3}(i\alpha\rho_c u_c' R_L)^{1/2}(y - y_c)^{3/2}} \right] \tag{120}$$

(iii) General characteristics of the curve of neutral stability

We now use the results of this section to discuss some general characteristics of the curve of neutral stability. For boundary-layer type of flows, the boundary conditions are

$$y = 0, \infty : \varphi = \varphi' = 0 \tag{121}$$

Noting that the required solution is given by a linear combination of $\tilde{\varphi}$ and $\hat{\varphi}$, the boundary condition at $y = 0$ leads to the characteristic-value equation

$$\frac{\tilde{\varphi}'(0)}{\tilde{\varphi}(0)} = \frac{\hat{\varphi}'(0)}{\hat{\varphi}(0)} \tag{122}$$

The asymptotic approximation to φ given by using the WKB method is

$$\hat{\varphi} \approx \left\{ e^{-\sqrt{\alpha R_L} \int_{y_c}^{y} \sqrt{i\rho(u-c)}\, dy} \right\} \times \times \left\{ (u - c)^{-5/4} + \ldots \right\} \tag{123}$$

Then

$$\frac{\hat{\varphi}'(y)}{\hat{\varphi}(y)} = -[i\rho\alpha R_L(u - c)]^{1/2} - \frac{5}{4}\frac{u'}{u - c} + \ldots \tag{124}$$

Note that since the two terms on the right-hand side above are comparable when

$$|y - y_c| = O(\rho\alpha R_L)^{-1/2} \tag{125}$$

the above approximation to $\hat{\varphi}$ cannot therefore give the correct asymptotic behaviour of the lower branch of the neutral-stability curve on which

$$(y - y_c) = O(\rho\alpha R_L)^{-1/2}$$

However, because of the simplicity of this approximation it is worth while to explore its merits in detail.

Noting that the WKB approximation to $\hat{\varphi}$ is valid only in the sector

$$-\frac{7\pi}{4} < \arg(y - y_c) < \frac{5\pi}{4}$$

(see next section), one finds

$$\frac{c}{u_0'}\frac{\hat{\varphi}'(0)}{\hat{\varphi}(0)} = -Z^{3/2}\,e^{-i\pi/4} + \frac{5}{4} \tag{126}$$

where

$$Z = \left(\frac{\rho_0\,\alpha R_L c^3}{u_0^2}\right)^{1/3}, \qquad u_0 = u(0)$$

Writing

$$W \equiv g + ih = \left[1 + \frac{u'\,\hat{\varphi}(0)}{c\,\hat{\varphi}'(0)}\right]^{-1} \tag{127}$$

equation (122') becomes

$$\frac{W-1}{W} = G(Z) \tag{128}$$

where

$$G(Z) = \left(Z^{3/2}\,e^{-i\pi/4} - \frac{5}{4}\right)^{-1}$$

or

$$Z^{3/2} = \left[\frac{1}{G(Z)} + \frac{5}{4}\right]e^{i\pi/4} \tag{129}$$

Also, note that

$$G(Z) = \begin{cases} -\dfrac{4}{5}\left[1 + \dfrac{4}{5}e^{-i\pi/4}\,Z^{3/2} + \ldots\right], & Z \to 0 \\ \dfrac{e^{i\pi/4}}{Z^{3/2}}\left[1 + \dfrac{5}{4}\dfrac{e^{i\pi/4}}{Z^{3/2}} + \ldots\ ,\right] & Z \to \infty \end{cases} \tag{130}$$

Some general stability characteristics in a viscous fluid can now be deduced by means of the asymptotic behaviour of the above expressions.

The inviscid solutions are (see Chapter IV)

$$\tilde{\varphi}_2(z) = 1 + \ldots + \left[\frac{T_c}{u_c'}\left(\frac{u'}{T}\right)'_{u=c}\tilde{\varphi}_1(z)\right]\left[\ln|z| - i\pi\right], \quad z<0 \tag{131}$$

$$\tilde{\varphi}_2(z) = 1 + \ldots + \left[\frac{T_c}{u_c'}\left(\frac{u'}{T}\right)'_{u=c}\tilde{\varphi}_1(z)\right]\left[\ln z\right], \quad z>0 \tag{132}$$

where

$$\tilde{\varphi}_1(z) = z + \frac{1}{2}\left(\frac{u_c''}{u_c'}\right) z^2 + \ldots, \quad z = y - y_c \tag{133}$$

One finds from equation (128) for real values of $c = c_s$,

$$\text{Im}\,[G(z)] = -\pi\,\frac{T_c}{c_s}\left(\frac{u'}{T}\right)'_{u=c_s} \tag{134}$$

From equation (129), one then obtains

$$\text{Im}\,[c] = \left(\frac{u_0'^2}{\rho_0 \alpha_s R_\text{L}}\right)^{1/3}\left[\frac{\pi T_c}{c_s}\left(\frac{u'}{T}\right)'_{u=c_s} + \frac{5}{4}\right] e^{i\pi/4} \tag{135}$$

If we restrict ourselves to cases where α and c do not approach zero along the neutral curve as $R_\text{L} \to \infty$, i.e. $\alpha = \alpha_s \neq 0$, $c = c_s \neq 0$, one finds from equation (135) for the case of an adiabatic wall

$$c_\text{i} > 0 \tag{136}$$

or that the disturbance $\alpha = \alpha_s$, neutrally stable ($c = c_s$) in the inviscid case, is unstable when viscosity comes into the picture. One therefore has the following viscous-stability criterion.

(a) *Viscous-stability criterion*

If a parallel compressible flow has an 'inviscid' neutral disturbance with $\alpha = \alpha_s \neq 0$, and $c = c_s \neq 0$, the disturbance $\alpha = \alpha_s$ is unstable in the real fluid for sufficiently large values of the Reynolds number, for the case of an adiabatic wall.

The counterpart of this criterion for the case of an incompressible fluid was established by Heisenberg. Thus for boundary-layer flows with an adiabatic wall, instability prevails for large Reynolds numbers, no matter whether the inviscid-instability criterion

$$\left(\frac{u'}{T}\right)'_{u=c} = 0$$

is satisfied or not.

Numerical work by Dunn and Lin (1955) and by Mack (1965) has showed the stabilising effect produced by cooling the wall, and the unstable modes produced by heating the wall.

A brief discussion of the significance of the neutral-stability curve (see Fig. I.2) is in order:

1. The region bounded by the two asymptotic branches of the curve signifies instability.

2. For free-shear-layer flows, the two branches of the curve approach two different asymptotes $\alpha = 0$ and $\alpha = \alpha_s$, leaving a finite unstable region for $R_L \to \infty$. For boundary-layer flows, the two branches approach the same asymptote $\alpha \to 0$, so that one has only the disturbance $\alpha = 0$ at $R_L \to \infty$.
3. One may conjecture that the neutral-stability curve could also correspond to cases of minimum damping with stable regions on both sides of it. In order to clarify this, we have to show that $c_i > 0$ in the neighbourhood of the neutral curve. Indeed, if we regard c as a function of the two independent parameters α and $\alpha R_L = R_L'$, we need only show that

$$\frac{\partial c_i}{\partial R_L'} < 0$$

for the upper branch of the neutral curve. Note that this is simply analogous to the viscous-stability criterion we have enunciated in the foregoing.

From equation (128),

$$G - 1 = -\frac{1}{W} \tag{137}$$

Regard, now, α as fixed and consider the variation of R_L' and Z with c. Taking the logarithmic derivatives of both sides of equation (128), one finds

$$\frac{G'}{G-1} = \frac{W'}{W} \tag{138}$$

Now the characteristic-value equation (138) involves the inviscid solutions only in the combination $\tilde{\varphi}'/\tilde{\varphi}$, and the latter can be directly found. It is also convenient to determine the asymptotic behaviour of the neutral curve. In the case of boundary-layer type flows, $\alpha \to 0$ along both branches of the neutral curve, as $\alpha R_L \to \infty$, and it is then natural to seek an approximation to W that is valid for small values of α.

Putting

$$H = \frac{\tilde{\varphi}'}{\tilde{\varphi}[T - M_1^2(u-c)^2]} \tag{139}$$

the inviscid equation

$$\left[\frac{\tilde{\varphi}'}{T - M_1^2(u-c)^2}\right]' - \left[\frac{\left\{\dfrac{u'}{T - M_1^2(u-c)^2}\right\}'}{u-c} + \frac{\alpha^2}{T}\right]\tilde{\varphi} = 0 \tag{140}$$

gives a Riccati equation

$$H' + [T - M_1^2(u-c)^2]H^2$$

$$= \left[\frac{\left\{ \dfrac{u'}{T - M_1^2(u-c)^2} \right\}'}{u-c} + \frac{\alpha^2}{T} \right] \tag{141}$$

Consider the boundary-layer flows for which the velocity profiles are monotonically increasing, and let

$$\frac{\tilde{\varphi}'}{\tilde{\varphi}} \to -\frac{\alpha}{\sqrt{T_1}} [T_1 - M_1^2(V_1 - c)^2]^{1/2} \quad \text{as } y \to \infty \tag{142}$$

Then from (139),

$$H(y) \to \frac{-\alpha}{\sqrt{T_1}\,[T_1 - M_1^2(V_1 - c)^2]^{1/2}}, \quad \text{as } y \to \infty \tag{143}$$

Seek a solution to equation (141) of the form

$$H(y) = \alpha \left[\frac{-1}{\sqrt{T_1}\,\{T_1 - M_1^2(V_1 - c)^2\}^{1/2}} \right]$$
$$+ \alpha^2 H_2(y) + \ldots \tag{144}$$

Upon substituting this expression into equation (141), and equating the coefficients of equal powers of α, one finds

$$H_2(y) = e^{-(2\alpha/\sqrt{T})y} \int_{\infty}^{y} \frac{\left\{ \dfrac{u'}{T - M_1^2(u-c)^2} \right\}'}{u-c} e^{(2\alpha/\sqrt{T})y}\, dy \tag{145}$$

so that from equations (127) and (139), one obtains

$$W = \left[1 + \frac{u_0'}{c} \frac{1}{T_1 H(0) \{T_0 - M_1^2 c^2\}} \right]^{-1}$$

or

$$W \approx 1 - \frac{u_0'}{c} \frac{1}{T_1 H(0) \{T_0 - M_1^2 c^2\}} \tag{146}$$

Then from equation (128), one obtains

$$G(Z) = 1 - \frac{(T_0 - M_1^2 c^2)\, c T_1 H(0)}{u_0'} +$$

$$+\left[\frac{(T_0 - M_1^2 c^2)\, c T_1 H(0)}{u_0'}\right]^2 + \ldots \tag{147}$$

Noting that on the upper branch of the neutral curve, $Z \to \infty$, and using (130) and (147), one obtains

$$\left(\frac{\partial c}{\partial R_L'}\right)_\alpha = \frac{-\dfrac{1}{2}\dfrac{e^{i\pi/4}}{Z^{3/2}}\dfrac{c}{R_L'}}{\left[\dfrac{u_0'/c}{1 + \dfrac{u_0'}{c}(T_0 - M_1^2 c^2)} + \dfrac{3}{2}\dfrac{e^{i\pi/4}}{Z^{3/2}}\right]} \tag{148}$$

From which for small c_i (i.e. for the neighbourhood of the neutral curve),

$$\left(\frac{\partial c_i}{\partial R_L'}\right)_\alpha < 0 \tag{149}$$

Note, however, that only for the particular disturbance $\alpha = 0$, for which one is on the lower branch of the neutral curve, a decrease of R_L promotes stability, so that for this particular case, viscosity is in the nature of stabilising the flow.

(5) ASYMPTOTIC PROPERTIES OF VISCOUS SOLUTIONS AND STOKES' PHENOMENON

The asymptotic properties of the regular solutions for $\eta \to \infty$ are expected to be comparable with the formal asymptotic solutions. It is the correlation between the two that enables one to find the correct determination of the path around the artificial singularity at $u = c$ introduced by the asymptotic methods.

The asymptotic behaviour of the Hankel functions $H_{1/3}^{(1)}\,[\frac{2}{3} i^{3/2} \zeta]$ and $H_{1/3}^{(2)}[\frac{2}{3} i^{3/2} \zeta]$ are (see Watson (1958))

$$H_{1/3}^{(1)}\left[\frac{2}{3} i^{3/2} \zeta\right] \sim \sqrt{\frac{2}{\pi}}\,(i^{3/2}\zeta)^{-1/2}\, e^{\frac{2}{3}\zeta}\, e^{5\pi i/4 - i(5\pi/12)} \times$$

$$\times\left[1 + O\left(\frac{1}{\zeta}\right)\right] \quad -\frac{7\pi}{4} < \arg\zeta < \frac{5\pi}{4} \tag{150}$$

$$H_{1/3}^{(2)}\left[\frac{2}{3} i^{3/2} \zeta\right] \sim \sqrt{\frac{2}{\pi}}\,(i^{3/2}\zeta)^{-1/2}\, e^{\frac{2}{3}\zeta}\, e^{\pi i/4} + i\frac{5\pi}{12} \times$$

$$\times\left[1 + O\left(\frac{1}{\zeta}\right)\right] - \frac{11\pi}{4} < \arg\zeta < \frac{\pi}{4} \tag{151}$$

The restrictions on the range of argument above imply that the path of integration is continued around the critical point y_c, from below the real axis in the complex y-plane when $u_c' > 0$, and is continued around from above when $u_c' < 0$.

Upon noting that the inviscid solutions must be valid asymptotic representations of equation (1)–(5) in a domain in which the asymptotic expansions of the viscous regular solutions for large values of R_L are valid, physically the above result implies that we view the inviscid neutral case as the limiting case of the self-excited case.

The restrictions on the range of argument in (150) and (151), however, do not preclude the existence of a connected region enclosing the points y_1, y_2, where the boundary conditions are to be imposed. None the less, the region of validity of the asymptotic expressions does not always include all the points of the real segment (y_1, y_2). In fact, the asymptotic solutions hold throughout the real interval (y_1, y_2) only when $c_i > 0$. For $c_i < 0$, there are two points on the real axis where the lines

$$\arg \zeta = -\frac{7\pi}{4}, \frac{5\pi}{4}$$

intersect the real axis (see Fig. 1), where the so-called Stokes' jump (as we shall see soon) of the asymptotic expansions occurs. These two points combine into one point when $c_i = 0$.

With the help of expressions (150) and (151) and using the legitimate process of integrating the asymptotic expansions term by term, one obtains

$$f_{11}^{(0)} \sim A\zeta^{-1/2}\, e^{\frac{2}{3}\zeta\, e^{5\pi i/4}} + \ldots \tag{152}$$

$$f_{12}^{(0)} \sim B\zeta^{-1/2}\, e^{\frac{2}{3}\zeta\, e^{\pi i/4}} + \ldots \tag{153}$$

The comparison between equations (152), (153) and (115) gives

$$\frac{2}{3}\zeta\, e^{5\pi i/4} \approx -\sqrt{\alpha R_L}\, \hat{G}(y)$$

or

$$\zeta \approx \left[\frac{3}{2}\int_{y_c}^{y} \sqrt{\alpha R_L\{\rho(u-c)\}}\; dy\right] \tag{154}$$

Note that this comparison has also enabled us to make a suitable choice for $\hat{g}_0(y)$ in the WKB solution (115).

(a) *Stokes' phenomenon*

Consider a path below the point y_c. Let $c_i = 0$ and $u_c' > 0$, then

$$f_{11}^{(0)} \sim A\,\zeta^{-1/2}\, e^{\frac{2}{3}\zeta\, e^{5\pi i/4}}, \quad \eta > 0 \tag{155}$$

$$f_{11}^{(0)} \sim A\,|\zeta|^{-1/2}\,e^{\frac{2}{3}|\zeta|\,e^{-\pi i/4}+3\pi i/4}, \quad \eta<0 \tag{156}$$

$$f_{12}^{(0)} \sim B\,\zeta^{-1/2}\,e^{\frac{2}{3}\zeta\,e^{\pi i/4}}, \quad \eta>0 \tag{157}$$

$$f_{12}^{(0)} \sim B\,|\zeta|^{-1/2}\,e^{\frac{2}{3}|\zeta|\,e^{-5\pi i/4}+3\pi i/4}, \quad \eta<0 \tag{158}$$

Whereas for a path above the point y_c one obtains

$$\bar{f}_{11}^{(0)} \sim A\,\zeta^{-1/2}\,e^{\frac{2}{3}\zeta\,e^{5\pi i/4}}, \quad \eta>0 \tag{159}$$

$$\bar{f}_{11}^{(0)} \sim A\,|\zeta|^{-1/2}\,e^{\frac{2}{3}|\zeta|\,e^{11\pi i/4}-3\pi i/4}, \quad \eta<0 \tag{160}$$

$$\bar{f}_{12}^{(0)} \sim B\,\zeta^{-1/2}\,e^{\frac{2}{3}\zeta\,e^{\pi i/4}}, \quad \eta>0 \tag{161}$$

$$\bar{f}_{12}^{(0)} \sim B\,|\zeta|^{-1/2}\,e^{\frac{2}{3}|\zeta|\,e^{7\pi i/4}-3\pi i/4}, \quad \eta<0 \tag{162}$$

Note that, if $A = B$, without loss of generality, one obtains

$$\bar{f}_{11}^{(0)} = -\,i\,f_{12}^{(0)}$$

$$\bar{f}_{12}^{(0)} = -\,i\,f_{11}^{(0)} \tag{163}$$

Therefore, if we consider $\bar{f}_{11}^{(0)}$, $\bar{f}_{12}^{(0)}$ as the proper determinations, we would have to make the following 'crossing substitution', corresponding to a passage from $\eta>0$ to $\eta<0$:

$$f_{11}^{(0)} \to i\,\bar{f}_{12}^{(0)}$$

$$f_{12}^{(0)} \to i\,\bar{f}_{11}^{(0)} \tag{164}$$

Such a discontinuous change in the asymptotic form of a continuous function is a typical instance of Stokes' phenomenon (Friedrichs, 1955; Jefreys, 1962). Physically this phase discontinuity experienced by the longitudinal-velocity perturbation f across the critical point $u = c$ is, for instance, at the heart of the energy transfer from a gas stream to the surface waves in a liquid layer adjacent to it.

Recall the WKB approximation

$$\hat{f}_1 = \left[e^{\pm\sqrt{\alpha R_L}\int_{y_c}^{y}\sqrt{i\rho(u-c)}\,dy}\right] \times$$

$$\times\,[(u-c)^{-3/4} + \ldots] \tag{165}$$

The lines

$$\mathrm{Re}\left[\int_{y_c}^{y}\sqrt{i\rho(u-c)}\,dy\right] = 0 \tag{166}$$

are called the Stokes' lines (Erdelyi, 1956). The Stokes' lines divide the complex y-plane into parts in which the viscous solutions have distinctive asymptotic forms which are interrelated to each other, as we saw in the foregoing, and in particular, the dominancy and subdominancy of a solution, as is evident from (165), changes as it is traced across such a line. If $y \to \infty$ along a Stokes' line, the exponential factors in both asymptotic solutions (115) remain bounded and bounded away from zero. If $y \to \infty$ along any other ray in the complex y-plane, the leading term in one of the solutions (115) increases exponentially.

Near the critical point $y = y_c$, expand $(u - c)$ in a Taylor's series:

$$u - c \approx u_c' \cdot (y - y_c)$$

then the Stokes' lines are given by

$$\mathrm{Re}\left[\int_{y_c}^{y} \sqrt{i\rho_c u_c' \cdot (y - y_c)}\, \mathrm{d}y\right] = 0$$

or

$$\mathrm{Re}\left[\sqrt{i\rho_c u_c'}\,(y - y_c)^{3/2}\right] = 0 \tag{167}$$

Thus there are three distinct Stokes' lines radiating outward from the point y_c, with angular spacings $2\pi/3$. They are located in the positions given by

$$\arg(y - y_c) = \left(\frac{\pi}{6}, \frac{5\pi}{6}, \frac{3\pi}{2}, \ldots\right) - \frac{1}{3}\arg(\rho_c u_c') \tag{168}$$

In the neutral case, u_c' is real, and if $u_c' > 0$, one obtains the geometry shown in Fig. 2. Thus if $c_i > 0$, the critical point y_c lies above the real axis, and the solutions to the inviscid disturbances are then regular and valid all along the real axis. In this case, the effects of viscosity and heat conductivity on the disturbances are negligible. However, if $c_i \leqslant 0$, the inviscid solutions cannot be uniformly-valid asymptotic approximations to a viscous regular solution all along the real axis of y (the only possible exception being the case with $(u'/T)'_{u=c} = 0$). If $c_i = 0$, as we saw in Chapter IV, there arises a logarithmic singularity in one of the inviscid solutions at the critical point y_c, and there is a thin layer at the critical point y_c in which the disturbances vary rapidly and in which therefore the viscous effects must be taken into account even in the inviscid limit. If $c_i < 0$, there are two such inner critical layers because the lines

$$\arg\left[(\rho_c u_c')^{1/3}\,(y - y_c)\right] = -\frac{7\pi}{4}, \frac{5\pi}{4}$$

which delimit the region of validity of the asymptotic expansions of the viscous regular solutions for large values of R_L intersect the real axis in two points.

Thus the two inner viscous layers for damped disturbances coalesce into one in the case of neutral disturbances and disappear altogether for self-excited disturbances. Consequently, for the self-excited disturbances, the viscous solutions tend to the corresponding inviscid solutions throughout the part of the real axis of the complex y-plane corresponding to the flow field, as $R_L \to \infty$. For the damped disturbances, the viscous solutions do not tend to the corresponding

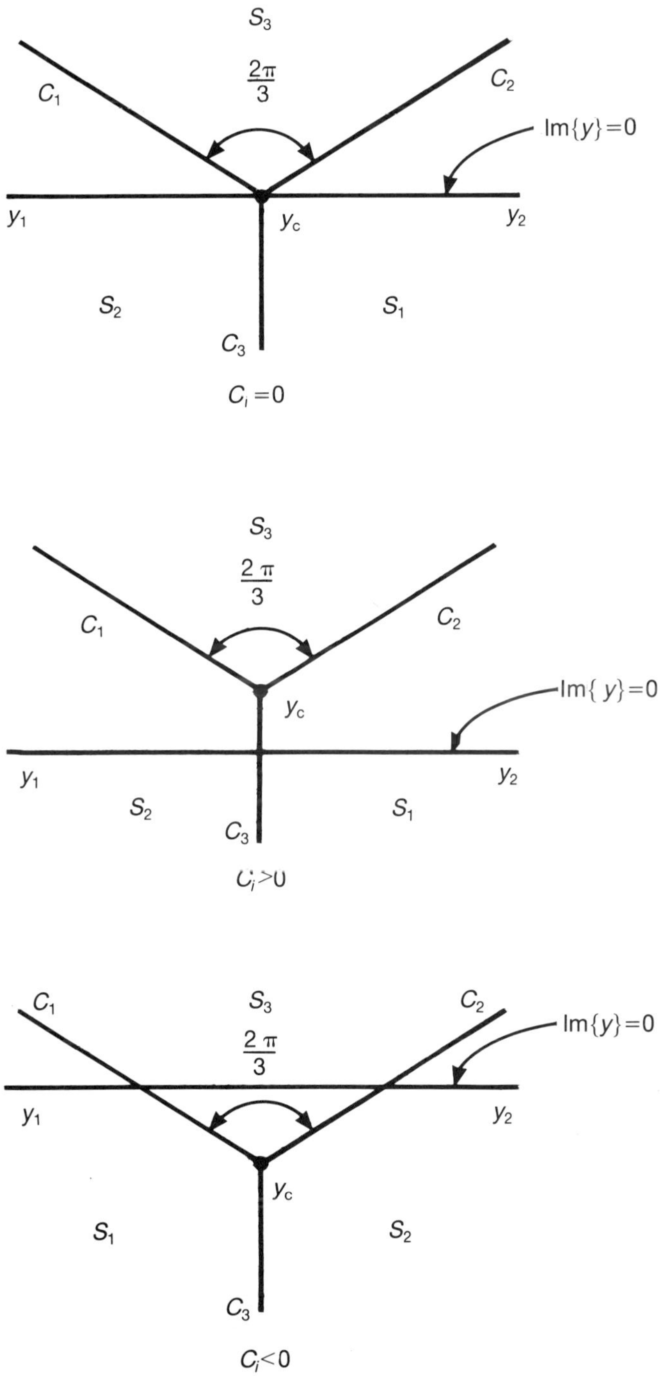
S3
2π/3
C1
C2
Im{y}=0
y1
yc
y2
S2
S1
C3
Ci=0
S3
2 π/3
C1
C2
Im{ y}=0
yc
y1
y2
S2
S1
C3
Ci>0
C1
S3
C2
Im{y}=0
2 π/3
y1
y2
yc
S1
S2
C3
Ci<0

Fig. 2

inviscid solutions along the whole of the real axis of the complex y-plane corresponding to the flow field.

The next problem then is to determine the proper transition of the asymptotic forms of the solutions across the Stokes' lines. Consider the behaviour of the solutions (115) in the various sectors of the complex y-plane. The solution $f_{11}(y)$ with a negative exponent is subdominant in the sector S_1 (and on the Stokes' lines C_2, C_3 it is neutral). Heading (1962) has shown that such a solution may be traced across the Stokes' lines C_2, C_3 into the sectors S_2, S_3 in which it is dominant up to but not including the Stokes' line C_1. Similarly, the solution $f_{12}(y)$ with a positive exponent is subdominant in S_2, and it can therefore be traced across C_1, C_3 into the sectors S_1, S_3 in which it is dominant up to but not including C_2. Both solutions are dominant in S_3.

A branch cut has to be inserted from y_c to ascertain the roots that appear in the solutions (115). The position of this cut can be chosen arbitrarily provided we ensure that the continuity, though not the form of a solution, is preserved as it is traced across the cut. Even this difficulty can be avoided if, following Olver (1959), the cut for f_{12} is placed along C_1 and the cut for f_{11} is placed along C_2.

(b) *Asymptotic properties of the inner solution*

Recall equation (34):

$$\frac{d^2 \hat{X}}{dy^2} + \lambda^2 (1 - y) \hat{X} = 0 \tag{34}$$

Put

$$Z = \lambda^{1/3} (1 - y) \tag{169}$$

so that equation (34) becomes

$$\frac{d^2 \hat{X}}{dZ^2} + Z \hat{X} = 0 \tag{170}$$

If,

$$G(\omega) = \int_{-\infty}^{\infty} \hat{X}(Z)\, e^{-i\omega Z} dZ \tag{171}$$

then the Fourier transfrom of equation (170) is

$$-\omega^2 G(\omega) + i \frac{dG(\omega)}{d\omega} = 0 \tag{172}$$

from which

$$G(\omega) = A\, e^{-i(\omega^3/3)} \tag{173}$$

where A is an arbitrary constant.

Thus,

$$\hat{X}(Z) = A \int_{-\infty}^{\infty} d\omega \, e^{i(\omega Z - \omega^3/3)} \tag{174}$$

which is an Airy integral.

We now use the saddle-point method to find the asymptotic forms of equation (172) corresponding to $Z \to \pm\infty$.

(i) $Z \to +\infty$: the saddle points occur on the real axis. Noting that we have an integral of the form

$$I(Z) = \int_{-\infty}^{\infty} e^{Zf(\omega)} d\omega, \quad Z \to +\infty \tag{175}$$

and near the saddle point $\omega = \omega_0$,

$$f(\omega) \approx f(\omega_0) + \frac{1}{2} f''(\omega_0) \cdot (\omega - \omega_0)^2 \tag{176}$$

and putting

$$f''(\omega_0) = \rho \, e^{i\psi}, \quad \omega - \omega_0 = s \, e^{i\phi} \tag{177}$$

one finds

$$I(Z) \approx e^{Zf(\omega_0)} \int_{-\infty}^{\infty} e^{-(Z/2)\rho s^2} e^{i\psi} ds$$

or

$$I(Z) \approx \sqrt{\frac{2\pi}{Z\rho}} \, e^{Zf(\omega_0) + i\phi} \tag{178}$$

where we have used the result

$$\int_{-\infty}^{\infty} e^{-as^2} ds = \sqrt{\frac{\pi}{a}} \tag{179}$$

Noting in our case that

$$f(\omega) = i\left(\omega - \frac{\omega^3}{3Z}\right)$$

$$f'(\omega) = i\left(1 - \frac{\omega^2}{Z}\right)$$

$$f''(\omega) = -\frac{2i\omega}{Z} \tag{180}$$

so that

$$f'(\omega_0) = 0 \quad \text{or} \quad \omega_0 = \pm\sqrt{Z}$$

$$f''(\omega_0) = \rho\, e^{i\psi} = \mp \frac{2i}{\sqrt{Z}} \quad \text{or} \quad \rho = \frac{2}{\sqrt{Z}}, \quad \psi = \mp \frac{\pi}{2}$$

$$\phi = -\frac{\pi}{4} \quad \text{(for the steepest descent)} \tag{181}$$

$$f(\omega_0) = \pm \frac{2}{3} i\sqrt{Z}$$

Thus,

$$\hat{X}(Z) \sim \sum_{+,-} \sqrt{\frac{2\pi\sqrt{Z}}{2Z}}\; e^{\pm\frac{2}{3} iZ^{3/2} \mp i\pi/4}$$

or

$$\hat{X}(Z) \sim \frac{2\sqrt{\pi}}{Z^{1/4}} \cos\left(\frac{2}{3} Z^{3/2} - \frac{\pi}{4}\right) \tag{182}$$

(ii) $Z \to -\infty$: the saddle points occur on the imaginary axis. Noting that now $(-Z) \to \infty$, so that

$$f(\omega) = -i\left(\omega - \frac{\omega^3}{3Z}\right)$$

$$f'(\omega) = -i\left(1 - \frac{\omega^2}{Z}\right)$$

$$f''(\omega) = \frac{2i\omega}{Z} \tag{183}$$

therefore

$$f'(\omega_0) = 0 \quad \text{or} \quad \omega_0 = \pm i\sqrt{-Z}$$

$$f''(\omega_0) = \rho\, e^{i\psi} = \pm \frac{2}{\sqrt{-Z}} \quad \text{or} \quad \rho = \frac{2}{\sqrt{-Z}}, \quad \psi = 0, \pi$$

$$\phi = 0 \quad \text{(for the steepest descent)} \tag{184}$$

$$f(\omega_0) = \pm \frac{2}{3}\sqrt{-Z}$$

Thus,

$$\hat{X}(Z) \sim \sqrt{\frac{2\pi\sqrt{-Z}}{2(-Z)}}\; e^{(-Z)\frac{2}{3}\sqrt{-Z}}$$

or

$$\hat{X}(Z) \sim \frac{\sqrt{\pi}}{(-Z)^{1/4}} \, e^{-\frac{2}{3}(-Z)^{3/2}} \tag{185}$$

Thus corresponding to (185) one finds

$$\hat{f}_1(Z) \sim (-Z)^{3/4} \, e^{-\frac{2}{3}(-Z)^{3/2}} \tag{186}$$

which is consistent with that in equation (152) obtained previously for the regular solution in the overlapping range of validity $\epsilon \ll |y - y_c| \ll 1$. This verifies that the simplified equations of section (2) have the same order of accuracy as the complete equations (1)–(5) for the inner viscous layer associated with the critical point $u = c$.

APPENDIX: UNIFORMLY-VALID ASYMPTOTIC SOLUTIONS FOR VISCOUS DISTURBANCES

The exact solutions for the disturbances for a general basic compressible flow have not been obtained. The situation of primary interest is when one of the parameters (say R_L) is large, and most of the solutions are constructed within the framework of this feature. Thus the commonly-known solutions are valid either in the immediate neighbourhood of the critical point $u = c$ or far away from the critical point.

For $\alpha R_L \gg 1$, the formal asymptotic solutions are

$$\varphi = \tilde{\varphi}^{(0)}(y) + (\alpha R_L)^{-1} \, \tilde{\varphi}^{(1)}(y) + \ldots \tag{A.1}$$

$$\varphi = \left[e^{-\alpha R_L \int_{y_c}^{y} \sqrt{i\rho(u-c)}\, dy} \right] \left[(u-c)^{-5/4} + \ldots \right] \tag{A.2}$$

The inviscid theory (Chapter IV) showed that one of the formal asymptotic solutions of the form (A.1) (denoted by $\tilde{\varphi}_1$) is regular, whereas another (denoted by $\tilde{\varphi}_2$) exhibits a logarithmic singularity at the critical point $u = c$. The singularity in this asymptotic solution and the asymptotic solution of the type (A.2) show that they can be valid only if the immediate neighbourhood of the critical point is excluded.

The asymptotic solution valid in the neighbourhood of the critical point is obtained by introducing a new variable

$$\eta = \frac{y - y_c}{\epsilon}, \qquad \epsilon = (\alpha R_L)^{-1/3} \ll 1 \tag{A.3}$$

and seeking a solution as a power series in ϵ of the form

$$\varphi(\eta, \epsilon) = \epsilon \, \chi_1(\eta) + \epsilon^2 \, \chi_2(\eta) + \ldots \tag{A.4}$$

which is valid for finite values of η.

The formal identification of (A.4) with (A.1) and (A.2) in the asymptotic limit (which is necessary from a heuristic point of view) raises mathematical difficulties in that it is not known *a priori* whether (A.1), (A.2) and (A.4) have overlapping domains of validity. Besides, in many cases in practice, one boundary corresponds to finite $(y - y_c)$, whereas the other corresponds to finite η. Then neither (A.1), (A.2), nor (A.4) is adequate for solving the characteristic-value problem. It is therefore desirable to develop uniformly-valid asymptotic solutions in a finite region containing the critical point, which we do here by using the method of multiple scales for a given isoenergetic basic compressible flow (Shivamoggi, 1977b).

Consider an isoenergetic basic compressible flow with $M_1 \ll 1$:

$$\begin{aligned} &u = y, \quad p = 1 \\ &T = 1 - \tfrac{1}{2}(\sigma - 1) M_1^2 y^2 \\ &\rho = 1 + \tfrac{1}{2}(\sigma + 1) M_1^2 y^2 \end{aligned} \tag{A.5}$$

Introduce two independent variables

$$z = \mathrm{i}(y - c), \quad \xi = (\alpha R_\mathrm{L})^{1/3} z \tag{A.6}$$

and note that a typical variable q is given by

$$q(y) = q(z, \xi; \epsilon), \quad \epsilon = (\alpha R_\mathrm{L})^{-1/3} \ll 1 \tag{A.7}$$

Then equations (1)–(5) become

$$\begin{aligned} &\mathrm{i}\Delta \frac{\partial \varphi}{\partial \xi} + \epsilon \left[c(\sigma - 1) M_1^2 \left(\xi \frac{\partial \varphi}{\partial \xi} + \varphi \right) + \mathrm{i}\Delta \left(\frac{\partial \varphi}{\partial z} + f \right) \right] + \\ &\quad + \epsilon^2 \left[\epsilon\gamma + (\sigma - 1) M_1^2 \left\{ -\frac{1}{2} \mathrm{i} \xi^2 \frac{\partial \varphi}{\partial \xi} + c\xi \left(\frac{\partial \varphi}{\partial z} + f \right) + \right. \right. \\ &\quad \left. \left. - \mathrm{i}\xi\varphi \right\} \right] + O(\epsilon^3) = 0 \end{aligned} \tag{A.8}$$

$$\begin{aligned} &\left[\Delta\varphi + \frac{\mathrm{i}\pi}{\sigma M_1^2} \right] + \epsilon \left[-\mathrm{i}(\sigma - 1) M_1^2 c\xi\varphi + \Delta\xi f + \frac{\partial^2 f}{\partial \xi^2} \right] + \\ &\quad + \epsilon^2 \left[(\sigma - 1) M_1^2 \left(-\frac{1}{2} \xi^2 \varphi - \mathrm{i}c\xi^2 f \right) + \right. \\ &\quad \left. + \frac{1}{3} \alpha^2 \frac{\partial \varphi}{\partial \xi} + 2 \frac{\partial^2 f}{\partial \xi \partial z} \right] + O(\epsilon^3) = 0 \end{aligned} \tag{A.9}$$

$$\frac{i\dfrac{\partial\pi}{\partial\xi}}{\sigma M_1^2}+\epsilon\left[\frac{i\dfrac{\partial\pi}{\partial z}}{\sigma M_1^2}\right]+\epsilon^2\alpha^2\left[\Delta\xi\varphi+\frac{4}{3}i\frac{\partial^2\varphi}{\partial\xi^2}\right]+$$

$$+O(\epsilon^3)=0 \tag{A.10}$$

$$i(\sigma-1)\frac{\partial\varphi}{\partial\xi}+\epsilon(\sigma-1)\left[-\Delta M_1^2 c\varphi+if\right]+$$

$$+\epsilon^2\left[\epsilon\theta\Delta-i(\sigma-1)M_1^2\,\xi\Delta\varphi+i(\sigma-1)^2M_1^4c\xi\alpha\varphi+\right.$$

$$\left.+\alpha\frac{\partial^2\varphi}{\partial\xi^2}\right]+O(\epsilon^3)=0 \tag{A.11}$$

$$[\pi+\delta\gamma+\Delta\theta]+\epsilon i(\sigma-1)M_1^2c(\ \ \xi\gamma+\xi\theta)+$$

$$+\epsilon^2\tfrac{1}{2}(\sigma-1)M_1^2\,\xi^2(\gamma+\theta)+O(\epsilon^3)=0 \tag{A.12}$$

where

$$\Delta=1+\tfrac{1}{2}(\sigma-1)M_1^2c^2$$

$$\delta=1-\tfrac{1}{2}(\sigma-1)M_1^2c^2$$

Now seek asymptotic equations for f, φ, π, γ, θ of the form

$$f(x,\xi;\epsilon)=\sum_{n=0}^{\infty}\epsilon^n f_n(z,\xi)$$

$$\varphi(z,\xi;\epsilon)=\sum_{n=0}^{\infty}\epsilon^{n+1}\varphi_n(z,\xi)$$

$$\pi(z,\xi;\epsilon)=\sum_{n=0}^{\infty}\epsilon^n\pi_n(z,\xi)$$

$$\gamma(z,\xi;\epsilon)=\sum_{n=0}^{\infty}\epsilon^n\gamma_n(z,\xi)$$

$$\theta(z,\xi;\epsilon)=\sum_{n=0}^{\infty}\epsilon^n\theta_n(z,\xi) \tag{A.13}$$

Substituting (A.13) into equations (A.8)–(A.12), and equating the coefficients of equal powers of ϵ, one obtains

$$\frac{\partial\varphi_0}{\partial\xi}+f_0=0 \tag{A.14}$$

$$\Delta\varphi_0 + \Delta\xi f_0 + \frac{\partial^2 f_0}{\partial\xi^2} = 0 \tag{A.15}$$

$$\frac{\partial\pi_0}{\partial\xi} = 0 \tag{A.16}$$

$$\xi\theta_0 + \frac{\sigma}{\Delta}\frac{\partial^2\theta_0}{\partial\xi^2} - (\sigma-1)M_1^2 c\varphi_0 = 0 \tag{A.17}$$

$$\delta\gamma_0 + \Delta\theta_0 = 0 \tag{A.18}$$

and

$$\begin{aligned} &\mathrm{i}\Delta\frac{\partial\varphi_1}{\partial\xi} + c\xi(\sigma-1)M_1^2\frac{\partial\varphi_0}{\partial\xi} + \mathrm{i}\Delta\frac{\partial\varphi_0}{\partial z} + \\ &\quad + (\sigma-1)M_1^2 c\varphi_0 + \mathrm{i}\Delta f_1 + \xi\gamma_0 + (\sigma-1)M_1^2 c\xi f_0 = 0 \end{aligned} \tag{A.19}$$

$$\begin{aligned} &\Delta\varphi_1 + \Delta\xi f_1 + \frac{\partial^2 f_1}{\partial\xi^2} - \mathrm{i}(\sigma-1)M_1^2 c\xi\varphi_0 + \\ &\quad - \mathrm{i}(\sigma-1)M_1^2 c\xi^2 f_0 + 2\frac{\partial^2 f_0}{\partial\xi\partial z} = 0 \end{aligned} \tag{A.20}$$

From equations (A.14)–(A.20), one obtains

$$\left(\frac{\partial^4}{\partial\xi^4} + \Delta\xi\frac{\partial^2}{\partial\xi^2}\right)\varphi_0 = 0 \tag{A.21}$$

$$\begin{aligned} &\left(\frac{\partial^4}{\partial\xi^4} + \Delta\xi\frac{\partial^2}{\partial\xi^2}\right)\varphi_1 = \mathrm{i}(\sigma-1)M_1^2 c\left[-(1-c)\varphi_0 + \right. \\ &\quad \left. + 2\xi\frac{\partial\varphi_0}{\partial\xi} + \xi^2\frac{\partial\varphi_0}{\partial\xi} + \xi^2\frac{\partial^2\varphi_0}{\partial\xi^2} - \frac{1}{\Delta}\frac{\partial^3\varphi_0}{\partial\xi^3}\right] + \\ &\quad - \Delta\frac{\partial\varphi_0}{\partial z} - \Delta\xi\frac{\partial^2\varphi_0}{\partial\xi\partial z} - 3\frac{\partial^3\varphi_0}{\partial\xi^2\partial z} + \\ &\quad + 2\mathrm{i}\xi\gamma_0 + \mathrm{i}\xi^2\frac{\partial\gamma_0}{\partial\xi} + \mathrm{i}\frac{1}{\Delta}\frac{\partial^3\gamma_0}{\partial\xi^3} + \frac{3}{\Delta}\frac{\partial^2\gamma_0}{\partial\xi^2} \end{aligned} \tag{A.22}$$

The solutions of equation (21) are integrals of Airy functions. Alternatively, they may be expressed as (see Jefreys (1962))

$$F_k(\zeta) = \int_{L_k}\frac{1}{t^2}\,\mathrm{e}^{(\zeta t + \frac{1}{3}t^3)}\mathrm{d}t \tag{A.23}$$

where

$$\zeta = \Delta^{-1/3}\xi$$

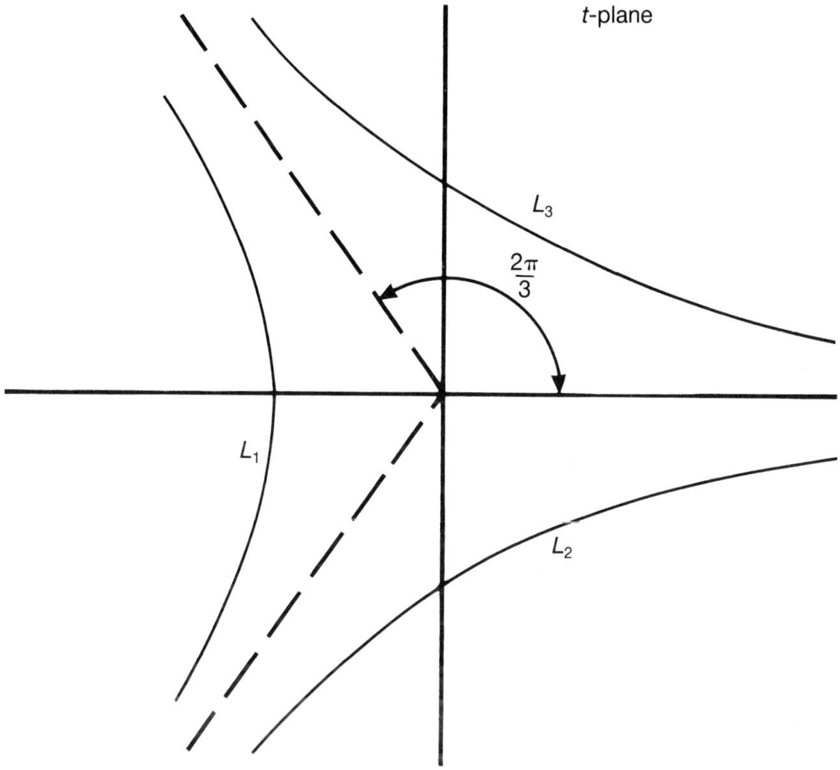

Fig. 3

and L_K is any one of the contours shown in Fig. 3. The solutions F_1, F_2, F_3 are linearly independent; however,

$$F_1 + F_2 + F_3 = F_0$$

where F_0 corresponds to a closed contour L_0 (not shown in the figure) around the origin. Thus, one writes

$$\varphi_0(z, \zeta) = A_0(z)\, F_1(\zeta) + B_0(z)\, F_2(\zeta) + C_0(z)\zeta + D_0(z) \qquad \text{(A.24)}$$

if the boundaries where the boundary conditions are to be imposed lies respectively in sectors S_1 and S_2 in Fig. 4. The functions A_0, B_0, C_0, D_0 are determined by imposing the condition that the asymptotic expansion (A.13) be uniformly valid, i.e. the terms of the latter must satisfy

$$\epsilon\,\varphi_n(z, \xi) = o(\varphi_{n-1}(z, \xi))$$

uniformly in z, as $\epsilon \to 0$, the same being true for their partial derivatives. This is accomplished by removing the secular terms in the next high-order equation, viz. (A.22).

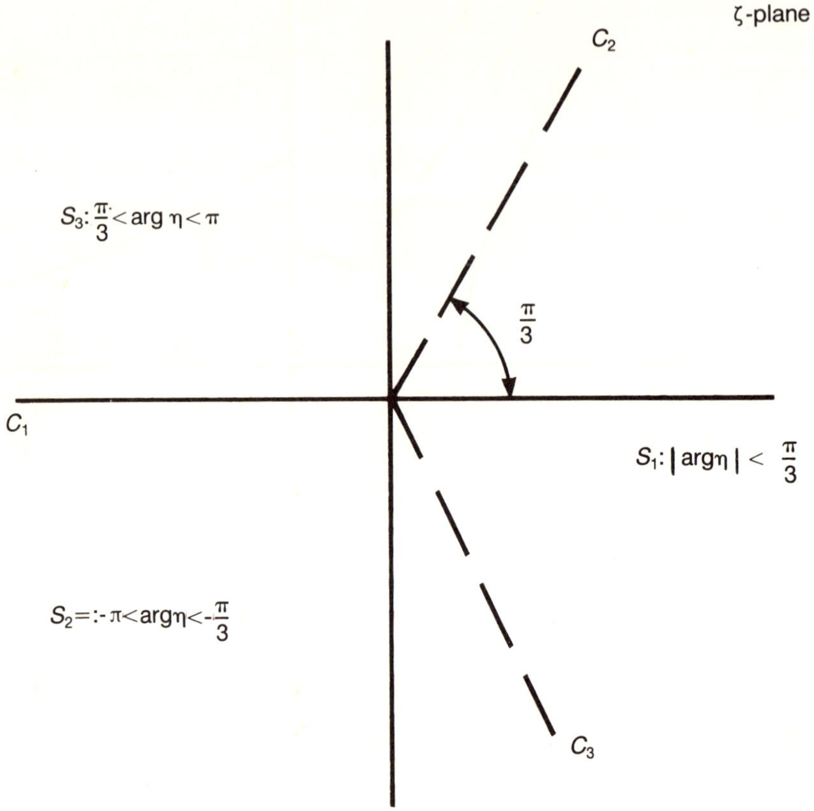

Fig. 4

Inspection of (A.22) using (A.24) shows that in order to remove the secular terms, one requires

$$A_0'(z) = B_0'(z) = C_0'(z) = 0$$

$$D_0'(z) + \mathrm{i}(\sigma - 1)\,\frac{M_1^2 c(1-c)}{\Delta}\,D_0(z) = 0 \tag{A.25}$$

or

$$A_0(z) = A_0, \quad B_0(z) = B_0, \quad C_0(z) = C_0$$

$$D_0(z) = \hat{D}_0 + D_0\,\mathrm{e}^{-\mathrm{i}(\sigma-1)M_1^2 c(1-c)z/\Delta} \tag{A.26}$$

Thus, one obtains a uniformly-valid solution

$$\varphi_0(z, \xi) = A_0\,F_1(\zeta) + B_0\,F_2(\zeta) + C_0\,\zeta + {}$$
$$+ \hat{D}_0 + D_0\,\mathrm{e}^{-\mathrm{i}(\sigma-1)M_1^2 c(1-c)z/\Delta} \tag{A.27}$$

9

Effects of Compressibility upon Stability Characteristics of a Free Shear Layer

INTRODUCTION

On a premise that on the large scale of long waves a generally smoothly-varying shear layer behaves like a vortex sheet with basic velocity

$$u(y) = \begin{cases} u(\infty), & y > 0 \\ u(-\infty), & y < 0 \end{cases}$$

the stability characteristics of a smoothly-varying shear-layer flow for long-wave disturbances are believed to correspond to those of a perturbed vortext sheet. On this account, the instability of a plane vortex sheet in a compressible fluid has been extensively investigated, e.g. by Landau (1944), Hatanaka (1949), and Miles (1958). In the apparent absence of a rigorous proof of this equivalence (indeed, Blumen *et al.* (1975) have warned of the dangers inherent in modelling real flows by a vortex sheet) we will treat smoothly-varying flow profiles subject to long-wave disturbances in an unbounded isoenergetic shear-layer flow, towards examining the effects of compressibility upon the inviscid-stability characteristics of a free shear flow (Shivamoggi, 1979). Within the framework of the linear theory, the compressibility effects are found to promote the stability of the perturbed basic flow. Next, in order to emphasise the trend of the compressibility effects revealed in the foregoing treatment that considered the flow field to be inviscid for the purpose of stability analysis, we will proceed to remove the latter restriction, and the viscous terms in the linearised-disturbance equations are now considered in the stability analysis (Shivamoggi, 1980) within the framework of an approximation procedure introduced by McKoen (1955) and Curle (1956). The resulting neutral-stability curve shows that the compressibility effects act toward promoting the stability of the perturbed basic flow thereby validating the conclusions reached in this direction previously under more restricted circumstances.

(1) COMPRESSIBILITY EFFECTS UPON INVISCID-STABILITY CHARACTERISTICS

One has for the inviscid problem

$$\frac{\mathrm{d}}{\mathrm{d}y}\left[\frac{\varphi'}{T-M_1^2(u-c)^2}\right]-\left[\frac{\dfrac{\mathrm{d}}{\mathrm{d}y}\left\{\dfrac{u'}{T-M_1^2(u-c)^2}\right\}}{u-c}+\frac{\alpha^2}{T}\right]\varphi=0 \tag{1}$$

$$y\to\pm\infty:\varphi\to 0 \tag{2}$$

The stability characteristics are deduced from the characteristic values of the problem (1), (2). Note that equation (1) has for $y\to\infty$, the asymptotic behaviour

$$\varphi\sim \mathrm{e}^{-\beta y},\quad \beta=\frac{\alpha}{T_\infty}\,[T_\infty-M_1^2\,W^2(\infty)],\quad W\equiv u-c \tag{3}$$

which, even for $\alpha \ll 1$, does not admit a power-series expansion in α, uniformly as $y\to\infty$. In order to avoid the nonuniformity of convergence at $y\to\infty$, seek solutions, for $\alpha\ll 1$, as power series in α of the form

$$\varphi_1\sim \mathrm{e}^{-\beta y}\sum_{n=0}^{\infty}\beta^n\chi_n(y,c),\quad 0\leqslant y<\infty \tag{4a}$$

and similarly,

$$\varphi_2\sim \mathrm{e}^{\beta y}\sum_{n=0}^{\infty}(-\beta)^n\,\psi_n(y,c),\quad 0\geqslant y>-\infty \tag{4b}$$

The boundary conditions at the interface $y=0$ are

$$\begin{aligned}\varphi_1(0^+)&=K\,\varphi_2(0^-)\\ \varphi_1'(0^+)&=K\,\varphi_2'(0^-)\end{aligned} \tag{5}$$

where K is a constant. Normalise φ_1,φ_2 according to

$$\begin{aligned}\chi_0(\infty)&=W(\infty)=W_\infty\\ \chi_n(\infty)&=0\\ \psi_0(-\infty)&=W(-\infty)=W_{-\infty}\\ \psi_n(-\infty)&=0\end{aligned} \tag{6}$$

One then has from equations (1) and (4a),

$$\left[\frac{W^2}{T-M_1^2 W^2}\left(\frac{\chi_0}{W}\right)'\right]' = 0$$

$$\left[\frac{W^2}{T-M_1^2 W^2}\left(\frac{\chi_{n+1}}{W}\right)'\right]' = \frac{2W\chi_n'}{T-M_1^2 W^2} + $$

$$-\frac{(T'-2M_1^2 WW')W\chi_n}{(T-M_1^2 W^2)^2} + $$

$$+\left[\frac{1}{T(T-M_1^2 W^2)} - \frac{1}{T-M_1^2 W^2}\right] W\chi_{n-1}, \quad n \geqslant 0 \tag{7}$$

so that

$$\chi_0(y) = W(y)$$

$$\chi_{n+1} = W\left[\int_\infty^y \frac{T-M_1^2 W^2}{W^2}\, dy_1 \times\right.$$

$$\times \int_\infty^{y_1}\left\{\frac{2W\chi_n'}{T \quad M_1^2 W^2} - \frac{(T'-2M_1^2 WW')W\chi_n}{(T-M_1^2 W^2)^2} + \right.$$

$$\left.\left.+\left[\frac{1}{T(T-M_1^2 W^2)} - \frac{1}{T-M_1^2 W^2}\right] W\chi_{n-1}\right\} dy_2\right] \tag{8}$$

Explicitly,

$$\chi_0(y) = W(y)$$

$$\chi_1(y) = W(y)\int_y^\infty \left[1 - \frac{W_\infty^2}{W^2}\left(\frac{T-M_1^2 W^2}{T_\infty - M_1^2 W_\infty^2}\right)\right] dy_1 \tag{9}$$

etc.

The equations for χ_n are the same as (7); the analogues of (8), (9) have ∞ replaced by $-\infty$ in the limits of integration, and W_∞ by $W_{-\infty}$.

The boundary conditions at $y = 0$ then give

$$\chi(0) = K\,\psi(0)$$

$$\chi'(0) - \beta\chi(0) = K\,[\psi'(0) + \beta\psi(0)] \tag{10}$$

From which one finds

$$\psi(0)\chi'(0) - \chi(0)\psi'(0) - 2\beta\psi(0)\chi(0) = 0 \tag{11}$$

which becomes for $M_1 \ll 1$, upon using (9),

$$-\beta\left[W_{\infty}^2\left\{\frac{T_0}{T_{\infty}}\left(1+\frac{M_1^2 W_{\infty}^2}{T_{\infty}}\right)\right\}+\right.$$

$$\left.+W_{-\infty}^2\left\{\frac{T_0}{T_{-\infty}}\left(1+\frac{M_1^2 W_{-\infty}^2}{T_{-\infty}}\right)\right\}\right]+O(\beta^2)=0 \qquad (12)$$

where $T_0 = T(0)$. From equation (12), there follows

$$\left[M_1^2\left(1+\frac{T_{\infty}^2}{T_{-\infty}^2}\right)\right]c^4+\left[4M_1^2\left(-u_{-\infty}-\frac{T_{\infty}^2}{T_{-\infty}^2}u_{-\infty}\right)\right]c^3+$$

$$+\left[T_{\infty}+\frac{T_{\infty}^2}{T_{-\infty}^2}T_{-\infty}+6M_1^2\left(u_{\infty}^2+\frac{T_{\infty}^2}{T_{-\infty}^2}u_{-\infty}^2\right)\right]c^2+$$

$$+\left[-2u_{\infty}T_{\infty}-2\frac{T_{\infty}^2}{T_{-\infty}^2}u_{-\infty}T_{-\infty}-4M_1^2\left(u_{\infty}^3+\frac{T_{\infty}^2}{T_{-\infty}^2}u_{-\infty}^3\right)\right]c+$$

$$+\left[u_{\infty}^2T_{\infty}+\frac{T_{\infty}^2}{T_{-\infty}^2}u_{-\infty}^2T_{-\infty}+M_1^2\left(u_{\infty}^4+\frac{T_{\infty}^2}{T_{-\infty}^2}u_{-\infty}^4\right)\right]=0$$

(13)

For an isoenergetic basic flow, one has $T \approx 1-[(\sigma-1)/2]\,M_1^2u^2$, so that one finds

$$\left[2-\left(\frac{\sigma-13}{2}\right)M_1^2(u_{\infty}^2+u_{-\infty}^2)\right]c^2+$$

$$-\left[2u_{\infty}+2u_{-\infty}-(\sigma-5)M_1^2(u_{\infty}^3+u_{-\infty}^3)\right]c+$$

$$+\left[u_{\infty}^2+u_{-\infty}^2-\left(\frac{\sigma-3}{2}\right)M_1^2(u_{\infty}^4+u_{-\infty}^4)\right]=0 \qquad (14)$$

From which corresponding to $c_i > 0$, one has

$$c=\frac{1}{2}\left[(u_{\infty}+u_{-\infty})-M_1^2\left\{\left(\frac{\gamma+3}{4}\right)(u_{\infty}^3+u_{-\infty}^3)+\right.\right.$$

$$\left.\left.-\left(\frac{\sigma-13}{4}\right)(u_{\infty}^2u_{-\infty}+u_{-\infty}^2u_{\infty})\right\}\right]+$$

$$+\frac{1}{2}\mathrm{i}\left[(u_\infty-u_{-\infty})-M_1^2\left\{(u_\infty+u_{-\infty})\times\right.\right.$$

$$\times\left[\left(\frac{\sigma+3}{4}\right)(u_\infty^3+u_{-\infty}^3)-\left(\frac{\sigma-13}{4}\right)(u_\infty^2u_{-\infty}+u_{-\infty}^2u_\infty)\right]$$

$$\left.\left.+\left(\frac{\sigma-7}{4}\right)(u_\infty^4+u_{-\infty}^4)+\left(\frac{\sigma-13}{4}\right)u_\infty^2u_{-\infty}^2\right\}\right] \tag{15}$$

which reduces in the limit $M_1 \to 0$ to the result deduced by Drazin and Howard (1962) who studied the same problem for the case of an incompressible fluid. It is seen that the effect of compressibility is to promote the stability of the perturbed basic flow. The stabilising nature of the compressibility effects on shear-layer flows was also found by Gill and Drazin (1965) and by Blumen (1970). Equation (15) also signifies that the shear-layer flows, for which $u_\infty = -u_{-\infty}$, are always unstable to long waves. Equation (15) further shows that for a shear-layer flow $c_r = 0$ – a result that *continues* to be valid even in the presence of compressibility effects. This result can also be established directly from equations (1), (2).

In order to prove this, note that if $u(y)$ is monotonically varying, then $(u - c_r)$ has only one zero, and therefore there is only one point where

$$(u'/T)'_{u=c_r} = 0$$

Since the number of unstable modes is the same as the number of points where

$$(u'/T)'_{u=c_r} = 0$$

there can be only one unstable mode under these circumstances.

THEOREM: *If $u(y)$ is an odd monotonic profile with only one point at the origin such that*

$$(u'/T)'_{u=c_r} = 0$$

then any unstable mode must have purely imaginary c.

Proof: If $\varphi(y)$ is a characteristic function of equation (7) with $c = c_r + \mathrm{i}c_i$ $(c_i > 0)$, then $\varphi(-y)$ is also a characteristic function with the same α, but with $c = -c_r + \mathrm{i}c_i$, since $u(y)$ is odd. Since from the foregoing, there can be only one unstable mode for a given α, one requires $c_r = 0$.

(2) COMPRESSIBILITY EFFECTS UPON VISCOUS-STABILITY CHARACTERISTICS

The viscous problem has the following governing equations:

$$\mathrm{i}(u-c)\gamma+\rho(\varphi'+\mathrm{i}f)+\rho'\varphi=0 \tag{16}$$

$$\alpha\rho\,[\mathrm{i}(u-c)f+u'\varphi]=-\frac{\mathrm{i}\alpha\pi}{\sigma M_1^2}+\frac{1}{R_\mathrm{L}}\,[f''+\alpha^2(\mathrm{i}\varphi'-2f)]+$$
$$-\frac{2}{3}\frac{\alpha^2}{R_\mathrm{L}}\,(-f+\mathrm{i}\varphi') \tag{17}$$

$$\alpha^2\rho\,[\mathrm{i}(u-c)\varphi]=-\frac{\pi'}{\sigma M_1^2}+\frac{\alpha}{R_\mathrm{L}}\,(2\varphi''+\mathrm{i}f'-\alpha^2\varphi)+$$
$$-\frac{2}{3}\frac{\alpha}{R_\mathrm{L}}\,(\varphi''+\mathrm{i}f') \tag{18}$$

$$\alpha\rho\,[\mathrm{i}(u-c)\theta+T'\varphi]=-\alpha\rho T(\sigma-1)\,(\varphi'+\mathrm{i}f)+\frac{\alpha}{R_\mathrm{L}}\,(\theta''+\alpha^2\theta)+$$
$$+\frac{\sigma(\sigma-1)M_1^2}{R_\mathrm{L}}\,[2u'(f'+\mathrm{i}\alpha^2\varphi)] \tag{19}$$

$$\frac{\pi}{p}=\frac{\gamma}{\rho}+\frac{\theta}{T} \tag{20}$$

where the nondimensional parameters are now defined by

$$R_\mathrm{L}=\frac{(V_1-V_2)L\rho_1}{2\mu_1},\quad M_1^2=\frac{(V_1-V_2)^2}{4\sigma RT_1}$$

where the subscripts 1 and 2 refer to the conditions in the high-speed stream and the low-speed stream, respectively.

In the following, we consider antisymmetrical velocity profiles, viz. that appropriate to the free laminar shear layer between two parallel streams. The method used to obtain the asymptotic form of the neutral-stability curve is an adaptation of that used to examine symmetrical mean profiles of flow of the jet – and wake-type – originally due to McKoen (1955) and refined later by Curle (1956). In this method the full viscous equations governing the linearised disturbances are considered, except that the highest-order derivatives are assumed to be of importance only in the inner viscous layer in the neighbourhood of the critical point $u = c$. The modifications are required because symmetrical and antisymmetrical disturbances cannot arise here separately. However, the relevant form of disturbance, which is a combination of these, degenerates into a symmetrical form for infinite Reynolds numbers at the upper end of the neutral curve and into an antisymmetrical form at the other end.

The general theory is then applied to the primary flow

$$u = \frac{V_1 + V_2}{2}\left[1 + \frac{V_1 - V_2}{V_1 + V_2}\tanh\left(\frac{y}{L}\right)\right]$$

$$T = -\frac{V_1 V_2}{2} + C_p\,\frac{T_1 V_2 - T_2 V_1}{V_2 - V_1} + $$

$$-\frac{1}{2C_p}\left(\frac{V_1 + V_2}{2}\right)^2\left[1 + \frac{V_1 - V_2}{V_1 + V_2}\tanh\left(\frac{y}{L}\right)\right]^2$$

where L characterises the width of the transition layer, say, the primary flow being assumed to be isoenergetic. The constant part $(V_1 + V_2)/2$ in equation (8) is unimportant in problems of stability, for it may be included in the wave velocity c of the disturbance since the disturbance equations are linear. Further this may change c_r but not c_i.

In the case of free-shear-layer flows, the conventional viscous solutions are rejected because they exponentially diverge at either of the boundaries at infinity. Hence somewhat-different methods have to be used for the treatment of problems of free shear layers. In particular, the viscous effects must be therefore introduced through the second-order approximation to the inviscid solutions

$$\varphi = \tilde{\varphi}^{(0)}(y) + \frac{1}{\alpha R_L}\,\tilde{\varphi}^{(1)}(y) + \ldots$$

Further, since in a boundary-layer flow, instability occurs only at high Reynolds numbers, the stability theory for such flows makes use of mathematical methods appropriate for high values of the Reynolds number. But in the case of free-shear-layer flows, instability occurs at low Reynolds numbers so that new methods are necessary to investigate the problem. The net effect of these is that the nature of the stability characteristics of a free-shear-layer flow, as we shall see in the following, differs considerably from that for a boundary-layer flow.

(3) THE STABILITY CHARACTERISTICS OF A FREE SHEAR LAYER

Eliminating the pressure disturbances one finds from equations (17) and (18),

$$f\,[\rho'(u-c) + \rho u'] + f'\left[\rho(u-c) - \frac{\mathrm{i}\alpha}{R_L}\right] +$$

$$+\, f''\left[\frac{\mathrm{i}}{R_L}\right] + \varphi\left[-\mathrm{i}\rho' u' - \mathrm{i}\rho u'' + \mathrm{i}\rho\alpha^2(u-c) + \right.$$

$$\left. + \frac{\alpha^3}{R_L}\right] + \varphi'\,[-\mathrm{i}\rho u'] + \varphi''\left[\frac{\alpha}{R_L}\right] = 0 \qquad (21)$$

One also has from equations (16)–(20),

$$f = -\frac{\mathrm{i}M_1^2(u-c)u'\varphi - \mathrm{i}T\varphi'}{T - M_1^2(u-c)^2} + O\left(\frac{1}{R_L}\right) \tag{22}$$

where the viscous terms contain higher-order derivatives. Thus equation (21) becomes, upon neglecting the third-order derivatives,

$$\left[\frac{\varphi'}{T-M_1^2(u-c)^2}\right]' - \left[\frac{\left\{\dfrac{u'}{T-M_1^2(u-c)^2}\right\}'}{u} + \frac{\alpha_s^2}{T}\right]\varphi = g(y) \tag{23}$$

where

$$\begin{aligned}
g(y) = \varphi\Bigg\{ & \frac{\mathrm{i}}{u-c_1}\Bigg[-\mathrm{i}\rho'u' - \mathrm{i}\rho u'' + \frac{\alpha^3}{R_L} + \\
& + \{\rho'(u-c) + \rho u'\}\left\{\frac{-\mathrm{i}M_1^2(u-c)u'}{T-M_1^2(u-c)^2}\right\} + \\
& + \left\{\rho(u-c) - \frac{\mathrm{i}\alpha}{R_L}\right\}\Bigg\{\frac{-\mathrm{i}M_1^2u'^2 - \mathrm{i}M_1^2(u-c)u''}{T-M_1^2(u-c)^2} + \\
& + \frac{\mathrm{i}M_1^2(u-c)u'[T' - 2M_1^2(u-c)u']}{[T-M_1^2(u-c)^2]^2}\Bigg\}\Bigg] + \\
& + \frac{c_1\left\{\dfrac{u'}{T-M_1^2(u-c)^2}\right\}'}{u(u-c_1)} + \left(\frac{\alpha^2}{T} - \frac{\alpha_s^2}{T}\right)\Bigg\} + \\
& + \frac{\varphi'}{(u-c_1)\{T-M_1^2(u-c)^2\}}\Bigg\{\Bigg[-\mathrm{i}\rho u' + \\
& + \{\rho'(u-c) + \rho u'\}\left\{\frac{\mathrm{i}T}{T-M_1^2(u-c)^2}\right\} + \\
& + \left\{\rho(u-c) - \frac{\mathrm{i}\alpha}{R_L}\right\}\left\{\frac{-2\mathrm{i}M_1^2(u-c)u' + \mathrm{i}T'}{T-M_1^2(u-c)^2}\right\}\Bigg]\Bigg\}
\end{aligned}$$

and

(i) $u - c_1 = u - c - \dfrac{2i\alpha}{R_L}\,[T - M_1^2(u-c)^2]$,

(ii) we have used knowledge of a solution of the inviscid equation, viz.

$$\left[\frac{\tilde{\varphi}'}{T - M_1^2(u-c)^2}\right]' - \left[\frac{\left\{\dfrac{u'}{T - M_1^2(u-c)^2}\right\}'}{u-c} + \frac{\alpha^2}{T}\right]\tilde{\varphi} = 0 \tag{24}$$

say,

$$\tilde{\varphi} = \tilde{\varphi}_s, \quad \alpha = \alpha_s, \quad c = c_s \equiv 0$$

Approximate equation (23), for the case $M_1 \ll 1$, by

$$\left[\frac{\varphi'}{T - M_1^2(u-c)^2}\right]' - \left[\frac{\left\{\dfrac{u'}{T - M_1^2(u-c)^2}\right\}'}{u} + \frac{\alpha_s^2}{T}\right]\varphi$$

$$= \varphi\left[\frac{\alpha^2}{T} - \frac{\alpha_s^2}{T} + \frac{c_1\left\{\dfrac{u'}{T - M_1^2(u-c)^2}\right\}'}{u(u-c_1)} + \right.$$

$$\left. + \frac{i\alpha^3}{R_L(u-c_1)}\right] \equiv g_1(y) \tag{25}$$

Equation (25) has a nontrivial solution if and only if the right-hand side is orthogonal to the corresponding homogeneous solution, i.e.

$$\int_{-\infty}^{\infty} g_1(y)\tilde{\varphi}_s(y) = 0 \tag{26}$$

The general solution of equation (25) is given by

$$\varphi = \tilde{\varphi}_s \int_{-\infty}^{y}\left[\frac{T - M_1^2(u-c)^2}{\tilde{\varphi}_s^{\,2}} \int_{-\infty}^{y} g_1(y_2)\tilde{\varphi}_s(y_2)\,\mathrm{d}y_2\right]\mathrm{d}y_1 + K\,\tilde{\varphi}_s \tag{27}$$

Choose as a first approximation

$$\varphi(y) \approx P\,\tilde{\varphi}_s(y) + Qu(y) \tag{28}$$

where for the moment we only specify that

$$\alpha = 0 : P = 0$$

$$\alpha = \alpha_s : Q = 0 \tag{29}$$

Note that $\varphi(0) = 0$ for the lower end of the neutral curve, but $\varphi'(0) = 0$ for the upper end of the neutral curve. Thus,

$$\varphi(0) = P\tilde{\varphi}_s(0) = K\,\tilde{\varphi}_s(0) \tag{30}$$

$$\frac{T(0) - M_1^2[u(0) - c]^2}{\tilde{\varphi}_s(0)} \int_{-\infty}^{0} \left(\frac{\alpha^2}{T} - \frac{\alpha_s^2}{T}\right) [P\tilde{\varphi}_s + Qu]\,\tilde{\varphi}_s\,dy$$

$$= \varphi'(0) = Qu'(0) \tag{31}$$

where we have made use of the fact that at the ends of the neutral curve,

$$g_1(y) \approx \tilde{g}_1(y) = \varphi\left(\frac{\alpha^2}{T} - \frac{\alpha_s^2}{T}\right) \tag{32}$$

Thus,

$$P = \frac{Q}{\displaystyle\int_{-\infty}^{0} (\alpha^2 - \alpha_s^2)\frac{\tilde{\varphi}_s^2}{T}\,dy}\left[\frac{\tilde{\varphi}_s(0)\,u'(0)}{T(0) - M_1^2\{u(0) - c\}^2} + \right.$$

$$\left. - \int_{-\infty}^{0} (\alpha^2 - \alpha_s^2)\frac{\tilde{\varphi}_s u}{T}\,dy\right] \tag{33}$$

Integrating the homogeneous part of equation (25) from $-\infty$ to 0, one finds

$$\int_{-\infty}^{0}\left[u\left\{\frac{\tilde{\varphi}_s'}{T - M_1^2(u-c)^2}\right\}' - \left\{\frac{u'}{T - M_1^2(u-c)^2}\right\}'\tilde{\varphi}_s\right] dy +$$

$$- \alpha_s^2 \int_{-\infty}^{0} \frac{u\tilde{\varphi}_s}{T}\,dy = 0$$

from which

$$\frac{u'(0)}{T(0) - M_1^2\{u(0) - c\}^2}\,\tilde{\varphi}_s(0) + \alpha_s^2 \int_{-\infty}^{0} \frac{u\tilde{\varphi}_s}{T}\,dy = 0 \tag{34}$$

so that equation (33) becomes

$$\frac{P}{-\alpha^2 \displaystyle\int_{-\infty}^{0} \frac{u\tilde{\varphi}_s}{T}\,dy} = \frac{Q}{\displaystyle\int_{-\infty}^{0} (\alpha^2 - \alpha_s^2)\frac{\tilde{\varphi}_s^2}{T}\,dy} \tag{35}$$

Now the characteristic-value relationship is determined from equation (26), which is

$$\int_{-\infty}^{\infty} (\alpha^2 - \alpha_s^2)\frac{\varphi\tilde{\varphi}_s}{T}\,dy +$$

$$+\int_{-\infty}^{\infty} \frac{c_1 \left\{ \dfrac{u'}{T - M_1^2 (u-c)^2} \right\}'}{u(u-c_1)} \varphi \tilde{\varphi}_s \, dy +$$

$$+\int_{-\infty}^{\infty} \frac{i\alpha^3}{R_L} \frac{\varphi \tilde{\varphi}_s}{u - c_1} \, dy = 0 \tag{36}$$

where φ is given by (28), so that

$$\int_{-\infty}^{\infty} (\alpha^2 - \alpha_s^2) \frac{\varphi \tilde{\varphi}_s}{T} \, dy + \oint_{-\infty}^{\infty} \frac{c_1 \left\{ \dfrac{u'}{T - M_1^2 (u-c)^2} \right\}'}{u(u-c_1)} \varphi \tilde{\varphi}_s \, dy +$$

$$+ \frac{i\alpha^3}{R_L} \int_{-\infty}^{\infty} \frac{\varphi \tilde{\varphi}_s}{u - c_1} \, dy$$

$$= -\left(\frac{\varphi \tilde{\varphi}_s}{u'} \right)^* \pi i \left[\left\{ \frac{u'}{T - M_1^2 (u-c)^2} \right\}'^* + \frac{i\alpha^3}{R_L} \right] \tag{37}$$

where $\oint$ means that we recognise only the Cauchy principal value for the singular integrals above, and the superscript * refers to the point where $u(y^*) = c_1$.

Consider the primary flow nondimensionalised with respect to $(V_1 - V_2)/2$, L, T_1, as before, so that

$$u = \tanh y$$

$$T = T_0 - \frac{\sigma - 1}{2} M_1^2 \tanh^2 y \tag{38}$$

Then the inviscid equation (24), with $c_s = 0$, becomes

$$\tilde{\varphi}_s'' - \tilde{\varphi}_s' \left[\frac{T' - 2M_1^2 u u'}{T - M_1^2 u^2} \right] +$$

$$- \tilde{\varphi}_s \left[\frac{u''}{u} - \frac{u'}{u} \frac{T' - 2M_1^2 u u'}{T - M_1^2 u^2} + \frac{\alpha_s^2}{T} (T - M_1^2 u^2) \right] = 0 \tag{39}$$

which in turn becomes for $\alpha_s = 1$, for the foreprescribed mean flow,

$$\tilde{\varphi}_s'' + \tilde{\varphi}_s' \left[\frac{\tanh y \operatorname{sech}^2 y}{-\tanh^2 y + \dfrac{2T_0}{(\sigma + 1) M_1^2}} \right] +$$

$$+\tilde{\varphi}_s\left[-1+2\,\mathrm{sech}^2\,y-\frac{\mathrm{sech}^4\,y}{-\tanh^2 y+\dfrac{2T_0}{(\sigma+1)M_1^2}}+\frac{M_1^2\tanh^2 y}{T_0-\dfrac{\sigma-1}{2}M_1^2\tanh^2 y}\right]=0 \tag{40}$$

We take as an initial approximation, corresponding to $M_1 \ll 1$,

$$\tilde{\varphi}_s \approx \mathrm{sech}\,y \tag{41}$$

so that from equation (28)

$$\varphi(y) \approx P\,\mathrm{sech}\,y + Q\tanh y \tag{42}$$

Then equation (37) becomes

$$\int_{-\infty}^{\infty}\left[\frac{\alpha^2-1}{T_0-\dfrac{\sigma-1}{2}M_1^2\tanh^2 y}\right][P\,\mathrm{sech}\,y+Q\tanh y]\,\mathrm{sech}\,y\,dy$$

$$+\oint_{-\infty}^{\infty}\frac{-2\,\mathrm{sech}^2 y\left(T_0-\dfrac{\sigma+1}{2}M_1^2\tanh^2 y\right)+(\sigma+1)M_1^2\,\mathrm{sech}^4\,y}{\left(T_0-\dfrac{\sigma+1}{2}M_1^2\tanh^2 y\right)(\tanh y_1-c_1)}\times$$

$$\times\; c_1\,\mathrm{sech}\,y\,[P\,\mathrm{sech}\,y+Q\tanh y]\,dy+$$

$$+\frac{i\alpha^3}{R_L}\oint_{-\infty}^{\infty}\frac{\mathrm{sech}\,y\,[P\,\mathrm{sech}\,y+Q\tanh y]}{\tanh y_1-c_1}\,dy$$

$$=-\frac{[P\,\mathrm{sech}\,y^*+Q\tanh y^*]}{\mathrm{sech}\,y^*}\,\pi i\times$$

$$\times\left[\frac{-2\tanh y^*\,\mathrm{sech}^2\,y^*\left(T_0-\dfrac{\sigma+1}{2}M_1^2\tanh^2 y^*\right)+(\sigma+1)M_1^2\tanh y^*\,\mathrm{sech}^4\,y^*}{\left(T_0-\dfrac{\sigma+1}{2}M_1^2\tanh^2 y^*\right)^2}+\frac{i\alpha^3}{R_L}\right]$$

$$= -\frac{P\sqrt{1-c_1^2}+Qc_1}{\sqrt{1-c_1^2}}\,\pi\mathrm{i}\times$$

$$\times\left[\frac{-2c_1(1-c_1^2)\left(T_0-\dfrac{\sigma+1}{2}M_1^2c_1^2\right)+ \\ +(\sigma+1)M_1^2c_1(1-c_1^2)^2}{\left(T_0-\dfrac{\sigma+1}{2}M_1^2c_1^2\right)^2}+\frac{\mathrm{i}\alpha^3}{R_L}\right] \qquad (43)$$

where

$$c_1 \approx c+\frac{2\mathrm{i}\alpha}{R_L}$$

Putting $\tanh y = Z$, equation (43) becomes

$$\frac{2P(\alpha^2-1)}{\sqrt{\left(\dfrac{\sigma-1}{2}\right)M_1^2T_0}}\tanh^{-1}\sqrt{\left(\frac{\sigma-1}{2}\right)M_1^2T_0}\;+$$

$$+\int_{-1}^{1}\frac{-2\left(-\dfrac{\sigma+1}{2}M_1^2Z^2+T_0\right)+(\sigma+1)M_1^2(1-Z^2)}{\left(-\dfrac{\sigma+1}{2}M_1^2Z^2+T_0\right)^2(Z-c_1)}\times$$

$$\times c_1\left[P(1-Z^2)+QZ\sqrt{1-Z^2}\,\right]\mathrm{d}Z+$$

$$+\frac{\mathrm{i}\alpha^3}{R_L}\left[P\oint_{-1}^{1}\frac{\mathrm{d}Z}{Z-c_1}+Q\oint_{-1}^{1}\frac{Z\,\mathrm{d}Z}{\sqrt{1-Z^2}\,(Z-c_1)}\right]$$

$$= -\frac{P\sqrt{1-c_1^2}+Qc_1}{\sqrt{1-c_1^2}}\,\pi\mathrm{i}\times$$

$$\times\left[\frac{-2c_1(1-c_1^2)\left(-\dfrac{\sigma+1}{2}M_1^2c_1^2+T_0\right)+ \\ -(\sigma+1)M_1^2c_1(1-c_1^2)^2}{\left(T_0-\dfrac{\sigma+1}{2}M_1^2c_1^2\right)}-\frac{\mathrm{i}\alpha^3}{R_L}\right] \qquad (44)$$

Note the integrals

$$1.\quad \oint_{-1}^{1} \frac{dZ}{Z-c_1} = [\ln(Z-c_1)]_{-1}^{c_1-\epsilon} + [\ln(Z-c_1)]_{c_1+\epsilon}^{1} = \ln\frac{1-c_1}{1+c_1} \tag{45}$$

$$2.\quad \oint_{-1}^{1} \frac{-2\left(-\dfrac{\sigma+1}{2}M_1^2Z^2+T_0\right)(1-Z^2)}{\left(T_0-\dfrac{\sigma+1}{2}M_1^2Z^2\right)^2(Z-c_1)}\,dZ$$

$$\approx -\frac{2}{T_0}\oint_{-1}^{1}\frac{1-Z^2}{Z-c_1}\,dZ, \quad \text{for } M_1 \ll 1$$

$$= -\frac{2}{T_0}\oint_{-1}^{1}\left[-Z-c_1+\frac{1-c_1^2}{Z-c_1}\right]dZ$$

$$= -\frac{2}{T_0}\left[-2c_1+(1-c_1^2)\ln\frac{1-c_1}{1+c_1}\right] \tag{46}$$

$$3.\quad \oint_{-1}^{1}\frac{Z\,dZ}{\sqrt{1-Z^2}\,(Z-c_1)}$$

which becomes, upon putting $Z = \cos x$, $c_1 = \cos x_1$,

$$\oint_{0}^{\pi}\frac{\cos x\,dx}{\cos x-\cos x_1}$$

$$= \pi + c_1\oint_{0}^{\pi}\frac{dx}{\cos x - c_1}$$

$$= \pi + \left[\frac{c_1}{\sqrt{1-c_1^2}}\ln\frac{1-c_1\cos x+\sqrt{1-c_1^2}\sin x}{\cos x - c_1}\right]_0^{x_1-\epsilon} +$$

$$+\left[\frac{c_1}{\sqrt{1-c_1^2}}\ln\frac{1-c_1\cos x+\sqrt{1-c_1^2}\sin x}{\cos x - c_1}\right]_{x_1+\epsilon}^{\pi}$$

$$= \pi \tag{47}$$

$$4. \quad \oint_{-1}^{1} \frac{-2\left(T_0 - \dfrac{\sigma+1}{2} M_1^2 Z^2\right) + (\sigma+1) M_1^2 (1 - Z^2)}{\left(T_0 - \dfrac{\sigma+1}{2} M_1^2 Z^2\right)^2 (Z - c_1)} \times$$

$$\times\ c_1\, Q Z \sqrt{1 - Z^2}\, \mathrm{d}Z$$

$$\approx Q c_1 \oint_{-1}^{1} \left[-2 \frac{Z\sqrt{1-Z^2}}{(Z - c_1)} \right] \mathrm{d}Z, \quad \text{for } M_1 \ll 1$$

Putting $Z = \cos x$, $c_1 = \cos x_1$, one obtains

$$-2 Q c_1 \int_0^{\pi} \frac{\cos x \sin^2 x}{\cos x - c_1}\, \mathrm{d}x$$

$$= -2 Q c_1 \int_0^{\pi} \left[-\cos^2 x - c_1 \cos x + 1 - c_1^2 + \frac{c_1(1 - c_1^2)}{\cos x - c_1} \right] \mathrm{d}x$$

$$= -2 Q c_1 \pi \left(\frac{1}{2} - c_1^2 \right) \tag{48}$$

Using equations (45)–(48), in equation (44), one obtains

$$\frac{2P(\alpha^2 - 1)}{\sqrt{\left(\dfrac{\sigma-1}{2}\right) M_1^2 T_0}} \tanh^{-1} \sqrt{\left(\frac{\sigma-1}{2}\right) M_1^2 T_0} - \frac{2 Q c_1 \pi}{T_0} \left(\frac{1}{2} - c_1^2 \right) +$$

$$- \frac{2}{T_0} P c_1 \left[(1 - c_1^2) \ln \frac{1 - c_1}{1 + c_1} - 2 c_1 \right]$$

$$+ P \frac{\mathrm{i}\alpha^3}{R_\mathrm{L}} \ln \frac{1 - c_1}{1 + c_1} + Q \frac{\mathrm{i}\alpha^3}{R_\mathrm{L}} \pi$$

$$= \frac{P\sqrt{1 - c_1^2} + Q c_1}{\sqrt{1 - c_1^2}} \pi \mathrm{i} \times$$

$$\times \left[\frac{2 c_1 (1 - c_1^2) \left(T_0 - \dfrac{\sigma+1}{2} M_1^2 c_1^2 \right) + -(\sigma+1) M_1^2 c_1^2 (1 - c_1^2)^2}{\left(T_0 - \dfrac{\sigma+1}{2} M_1^2 c_1^2 \right)^2} - \frac{\mathrm{i}\alpha^3}{R_\mathrm{L}} \right] \tag{49}$$

Next using equations (38), and (41), equation (33) gives

$$\frac{P}{\alpha^2 \int_{-\infty}^{0} \dfrac{\tanh y \operatorname{sech} y \, dy}{\left(T_0 - \dfrac{\sigma - 1}{2} M_1^2 \tanh^2 y\right)}}$$

$$= \frac{Q}{(\alpha^2 - 1) \int_{-\infty}^{0} \dfrac{\operatorname{sech}^2 y \, dy}{\left(T_0 - \dfrac{\sigma - 1}{2} M_1^2 \tanh^2 y\right)}}$$

or

$$\frac{P}{\dfrac{2\alpha^2}{(\sigma - 1)M_1^2} \dfrac{1}{\sqrt{\dfrac{2}{(\sigma - 1)M_1^2}\left(T_0 - \dfrac{\sigma - 1}{2} M_1^2\right)}} \tan^{-1} \left\{ \dfrac{1}{\sqrt{\dfrac{2}{(\sigma - 1)M_1^2}\left(T_0 - \dfrac{\sigma - 1}{2} M_1^2\right)}} \right\}}$$

$$= \frac{Q}{\dfrac{2(\alpha^2 - 1)}{(\sigma - 1)M_1^2} \dfrac{1}{\sqrt{\dfrac{2T_0}{(\sigma - 1)M_1^2}}} \tanh^{-1} \dfrac{1}{\sqrt{\dfrac{2T_0}{(\sigma - 1)M_1^2}}}} \qquad (50)$$

Using equation (50) in equation (49), one obtains

$$\frac{2(\alpha^2 - 1)}{\sqrt{\left(\dfrac{\sigma - 1}{2}\right) M_1^2 T_0}} \tanh^{-1} \sqrt{\left(\frac{\sigma - 1}{2}\right) \frac{M_1^2}{T_0}} +$$

$$- \frac{2\pi c_1}{T_0} \left(\frac{1}{2} - c_1^2\right) \frac{\alpha^2 - 1}{\alpha^2} \sqrt{\left(1 - \frac{\sigma - 1}{2} \frac{M_1^2}{T_0}\right)} \times$$

$$\times \frac{\tanh^{-1} \dfrac{1}{\sqrt{2T_0/(\sigma-1)M_1^2}}}{\tan^{-1} \dfrac{1}{\sqrt{\dfrac{2}{(\sigma-1)M_1^2}\left(T_0 - \dfrac{\sigma-1}{2} M_1^2\right)}}} +$$

$$- \frac{2}{T_0} c_1 \left[-2c_1 + (1-c_1^2) \ln \frac{1-c_1}{1+c_1} \right]$$

$$+ \frac{i\alpha^3}{R_L} \ln \frac{1-c_1}{1+c_1} + \frac{i\alpha^3 \pi}{R_L} \frac{\alpha^2 - 1}{\alpha^2} \sqrt{1 - \frac{\sigma-1}{2} \frac{M_1^2}{T_0}} \times$$

$$\times \frac{\tanh^{-1} \dfrac{1}{\sqrt{2T_0/(\sigma-1)M_1^2}}}{\tan^{-1} \dfrac{1}{\sqrt{\dfrac{2}{(\sigma-1)M_1^2}\left(T_0 - \dfrac{\sigma-1}{2} M_1^2\right)}}}$$

$$= \frac{1}{\sqrt{1-c_1^2}} \left[\sqrt{1-c_1^2} + c_1 \frac{\alpha^2-1}{\alpha^2} \sqrt{\left(1 - \frac{\sigma-1}{2} \frac{M_1^2}{T_0}\right)} \times \right.$$

$$\left. \times \frac{\tanh^{-1} \dfrac{1}{\sqrt{2T_0/(\sigma-1)M_1^2}}}{\tan^{-1} \dfrac{1}{\sqrt{\dfrac{2}{(\sigma-1)M_1^2}\left(T_0 - \dfrac{\sigma-1}{2} M_1^2\right)}}} \right] \pi i \times$$

$$\times\left[\frac{\begin{array}{c}2c_1(1-c_1^2)\left(T_0-\dfrac{\sigma+1}{2}M_1^2c_1^2\right)+\\-(\sigma+1)M_1^2c_1^2(1-c_1^2)^2\end{array}}{\left(T_0-\dfrac{\sigma+1}{2}M_1^2c_1^2\right)^2}-\frac{i\alpha^3}{R_L}\right] \tag{51}$$

This equation is to be solved to give the relationship between α, c, and R_L. Noting that

$$1/R_L \ll 1$$

$$c \ll 1$$

at both ends of the neutral curve, from equation (51) one finds

$$2\alpha^2(\alpha^2-1)\left(1+\frac{\sigma-1}{2}M_1^2T_0\right)\frac{1}{T_0}+$$

$$+\frac{\pi c_1}{T_0}(1-\alpha^2)-\frac{i\alpha^3}{R_L}\pi(1-\alpha^2)$$

$$=\alpha^2\pi i\left[2c_1\left\{1-\frac{\sigma+1}{2}\frac{M_1^2}{T_0}\right\}\frac{1}{T_0}-\frac{i\alpha^3}{R_L}\right]\times$$

$$\times\left(1+\frac{\sigma-1}{2}\frac{M_1^2}{T_0}\right) \tag{52}$$

Equation (52) is then split into real and imaginary parts

$$2\alpha^2(\alpha^2-1)\left(1+\frac{\sigma-1}{2}\frac{M_1^2}{T_0}\right)\frac{1}{T_0}+\frac{\pi c_r}{T_0}(1-\alpha^2)$$

$$=\alpha^2\pi\left[\frac{\alpha^3}{R_L}-\frac{4\alpha}{R_L}\left(1-\frac{\sigma+1}{2}\frac{M_1^2}{T_0}\right)\frac{1}{T_0}\right]\times$$

$$\times\left(1+\frac{\sigma-1}{2}\frac{M_1^2}{T_0}\right) \tag{53}$$

and

$$\frac{2\pi\alpha}{R_LT_0}(1-\alpha^2)-\frac{\pi\alpha^3}{R_L}(1-\alpha^2)$$

$$= 2\alpha^2 \pi c_r \left(1 - \frac{\sigma + 1}{2} \frac{M_1^2}{T_0}\right) \left(1 + \frac{\sigma - 1}{2} \frac{M_1^2}{T_0}\right) \tag{54}$$

Eliminating c_r, one obtains

$$R_L = \frac{\dfrac{\pi\left[2 - 4\alpha^2 + 2\alpha^4 + T_0\left(2\alpha^4 - \alpha^6\right)\right]}{\left(1 + \frac{\sigma - 1}{2}\frac{M_1^2}{T_0}\right)\left(1 - \frac{\sigma + 1}{2}\frac{M_1^2}{T_0}\right)} + \pi\left[-2\alpha^6 + 8\alpha^4\left(1 - \frac{\sigma + 1}{2}\frac{M_1^2}{T_0}\right)\frac{1}{T_0}\right] \times \left(1 + \frac{\sigma - 1}{2}\frac{M_1^2}{T_0}\right)}{4\alpha^3(1 - \alpha^2)\left(1 + \frac{\sigma - 1}{2}\frac{M_1^2}{T_0}\right)\frac{1}{T_0}} \tag{55}$$

which in the incompressible limit $M_1 \to 0$ agrees with that obtained by Curle (1956). Equation (55) gives the neutral-stability curve as depicted in Fig. 1.

Fig. 1 shows that the effect of compressibility is to promote the stability of the perturbed basic flow. This implies that the conclusions reached in the previous section in this direction under more restricted circumstances, viz. within the framework of an inviscid model for the basic flow, continue to be valid under more real circumstances.

For the sake of comparison, we have superposed on Fig. 1 the neutral-stability curve for the case of a boundary layer in an incompressible fluid as obtained by Lin (1945). A few more observations of Fig. 1 are in order:

(i) the neutral-stability curves for free-shear layer flows extend to much lower Reynolds numbers than those for boundary-layer flows;
(ii) the free-shear layer flow is unstable for some disturbances at infinite Reynolds numbers, whereas the boundary-layer flow is unstable in that limit (this has led to a conjecture that the major instability mechanism in a free-shear-layer flow is inviscid in character, that does not operate in the boundary-layer flows);
(iii) for a free-shear layer flow the range of the wave number corresponding to instability decreases monotonically with decreasing Reynolds number, whereas a somewhat opposite situation prevails for a boundary-layer flow.

Finally it remains to be pointed out that in view of the effect of departures from the parallel-flow model arising at very low Reynolds numbers, the complete authenticity of the present results is not guaranteed in this range.

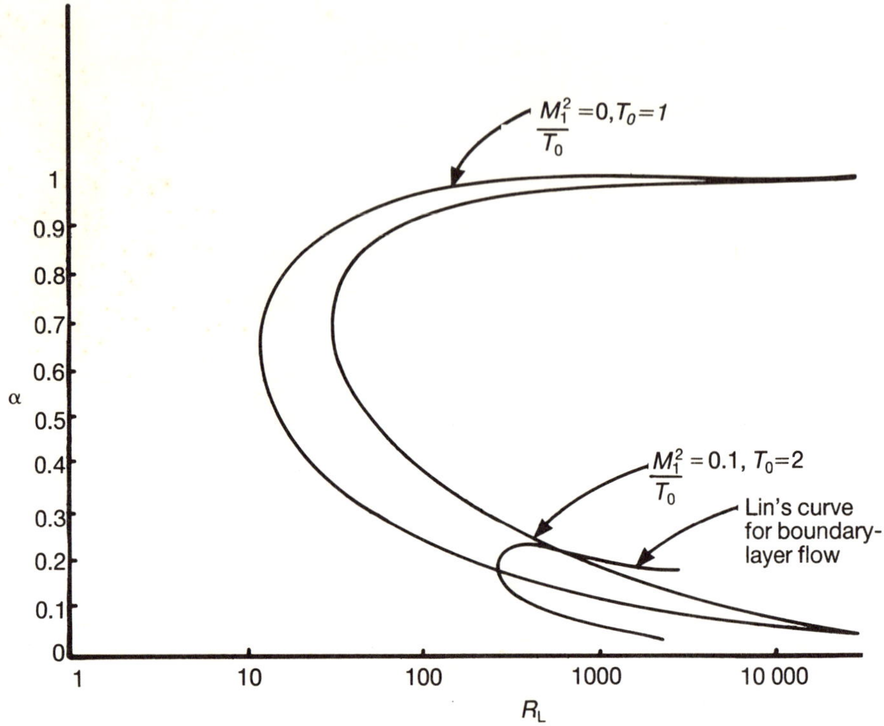

$\frac{M_1^2}{T_0}=0, T_0=1$
$\frac{M_1^2}{T_0}=0.1, T_0=2$
Lin's curve for boundary-layer flow
α
1
0.9
0.8
0.7
0.6
0.5
0.4
0.3
0.2
0.1
0
1
10
100
1000
10 000
R_L

Fig. 1

References

Blumen, W. (1970). *J. Fluid Mech.* **40**, 769.

Blumen, W., Drazin, P. G. and Billings, D. F. (1975). *J. Fluid Mech.* **71**, 305.

Butenin, N.W. (1965). *Theory of Nonlinear Oscillations*, Blaisdell Publishing Co.

Case, K.M. (1960). *Phys. Fluids* **3**, 143.

Curle, N. (1956). Aero. Res. Council 18426, FM2400.

Davey, A. (1973). *Q. J. Mech. Appl. Math.* **26**, 401.

Davies, S.J. and White, C.M. (1928). *Proc. Roy. Soc. (London).* **A119**, 92.

Dikii, L.A. (1961). *Sov. Phys. Dokl.* **5**, 1179.

Drazin, P.G. and Howard, L.N. (1962). *J. Fluid Mech.* **14**, 257.

Drazin, P.G. and Howard, L.N. (1966). Hydrodynamic Stability of Parallel Flow of Inviscid Fluid, in *Adv. Appl. Mech.*, Ed. G. Kuerti, 7, 1.

Drazin, P.G. and Reid, W.H. (1981). *Hydrodynamic Stability*, Cambridge Univ. Press.

Dunn, D.W. and Lin, C.C. (1955). *J. Aero. Sci.* **22**, 455.

Erdelyi, A. (1956). *Asymptotic Expansions*, Dover.

Fjortoft, T. (1950). *Geophys. Publ.* **17**, 1.

Friedrichs, K.O. (1942). *Fluid Dynamics*, Brown Univ., Providence.

Friedrichs, K.O. (1955). *Bull. Am. Math. Soc.* **61**, 485.

Gaster, M. (1962). *J. Fluid Mech.* **14**, 222.

Gill, A.G. & Drazin, P.G. (1965). *J. Fluid Mech.* **22**, 415.

Hall, P. and Smith F.T. (1982). *Studies Appl. Math.* **66**, 241.

Hatanaka, H. (1949). *J. Soc. Sci. Culture Japan* **2**, 3.

Heading J. (1962a). *Q. J. Mech. Appl. Math.* **15**, 215.
Heading, J. (1962). *Phase Integral Methods*, Methuen.
Heisenberg, W. (1924). *Ann. d'Phys.* **74**, 577.
Helmholtz, H. (1868). *Phil. Mag. (Transl.) (4)* **36**, 337.
Holstein, G. (1950). *Z. Angew. Math. Mech.* **30**, 25.
Horn, J. (1899). *Math. Ann.* **52**, 271.
Howard, L.N. (1961). *J. Fluid Mech.* **10**, 509.
Howard, L.N. (1964). *J. de Mecanique* **3**, 433.
Jefreys, H. (1924). *Proc. Roy. Soc. (London)* **A107**, 189.
Jefreys, H. (1925). *Proc. Roy. Soc. (London)* **A108**, 241.
Jefreys, H. (1962). *Asymptotic Expansions*, Clarendon Press.
Kelvin, Lord (1887). *Phil. Mag. (5)* **23**, 459.
Kemble, E.C. (1958). *Principles of Quantum Mechanics*, Dover.
Lakin, W.D., Ng, B.S. and Reid, W.H. (1978). *Phil. Trans. Roy. Soc. (London)* **289**, 347.
Landau, L.D. (1944). *Sov. Phys. Dokl.* **44**, 139.
Landau, L.D. (1946). *J. Phys. USSR* **10**, 25.
Langer, R.E. (1955). *Duke Math. J.* **22**, 525.
Laufer, J. and Vrebalovich, T. (1960). *J. Fluid Mech.* **9**, 257.
Lees, L. and Lin, C.C. (1946). NACA Tech. Note 1115.
Lin, C.C. (1945). *Q. Appl. Math.* **3**, 117, 218 and 277.
Lin, C.C. (1953). NACA Tech. Note 2887.
Lin, C.C. (1955). *Theory of Hydrodynamic Stability*, Cambridge Univ. Press.
Lock, R.C. (1954). *Proc. Cambr. Phil. Soc.* **50**, 105.
McKoen, C.H. (1955). Aero. Res. Council 17952, FM2313.
Mack, L. (1965). AGARDograph **27**, pt. I, 329.
Meksyn, D. and Stuart, J.T. (1951). *Proc. Roy. Soc. (London)* **A208**, 517.
Miles, J.W. (1957). *J. Fluid Mech.* **3**, 185.
Miles, J.W. (1958). *J. Fluid Mech.* **4**, 538.
Ng, B.S. (1976). *Q. Appl. Math.* **33**, 319.
Nishioka, M., Iida, S. and Ichikawa, Y. (1975). *J. Fluid Mech.* **72**, 731.
Olver, F.W.J. (1959). *J. Res. Natl. Bur. Std. (V.S.A.)* **B63**, 131.
Orr, W.M.F. (1907). *Proc. Roy. Irish Acad.* **A27**, 9.
Orszag, S.A. (1971). *J. Fluid Mech.* **50**, 689.
Pekeris, C.L. & Shkoller, B. (1967). *J. Fluid Mech.* **29**, 31.
Prandtl, L. (1935). Mechanics of Viscous Fluids, in *Aerodynamic Theory*, Ed. W.F. Durand, Vol. 3, Dover.
Rayleigh, Lord (1880). *Proc. London Math. Soc.* **11**, 57.
Rayleigh, Lord (1883). *Proc. London Math. Soc.* **14**, 170.
Rayleigh, Lord (1892). *Phil. Mag. (5)* **34**, 59.
Rayleigh, Lord (1912). *Proc. Roy. Soc. (London)* **A86**, 208.
Rayleigh, Lord (1913). *Phil. Mag. (6)* **26**, 776.
Rayleigh, Lord (1914). *Phil. Mag. (6)* **28**, 609.
Reid, W. H. (1965). Stability of Parallel Flows, in *Basic Developments in Fluid Dynamics*, Ed. M. Holt, Vol. 1, Academic Press.

Reynolds, O. (1883). *Phil. Trans. Roy. Soc. (London)* **174**, 935.
Reynolds, W.C. and Potter, M.C. (1967). *J. Fluid Mech.* **27**, 465.
Saric, W.S. and Nayfeh (1975). *Phys. Fluids* **18**, 945.
Schlichting, H. (1933). Gottingen Nachrichten, Math. Phys. K1., p. 181.
Schubauer, G.B. and Skramstad, H.K. (1947). *J. Aero. Sci.* **14**, 69.
Shen, S.F. (1954). *J. Aero. Sci.* **21**, 62.
Shivamoggi, B.K. (1977a). *J. de Mecanique* **16**, 227.
Shivamoggi, B.K. (1977b). *J. Fluid Mech.* **79**, 745.
Shivamoggi, B.K. (1978a). *Proc. Indian Acad. Sci.* **87A**, 107.
Shivamoggi, B.K. (1978b). *Acta Mech.* **30**, 197.
Shivamoggi, B.K. (1979). *Z. Angew. Math. Mech.* **59**, 405.
Shivamoggi, B.K. (1980). *Mec. Appliquee* **25**, 353.
Shivamoggi, B.K. (1982). *Acta Mech.* **44**, 137.
Sommerfeld, A. (1908). *Proc. IV Intr. Congr. Math. Rome,* 116.
Squire, H.B. (1933). *Proc. Roy. Soc. (London)* **A142**, 621.
Stuart, J.T. (1958). *J. Fluid Mech.* **4**, 1.
Stuart, J.T. (1960). *J. Fluid Mech.* **9**, 353.
Synge, J.L. (1933). *Trans. Roy. Soc. Can. (3)* **27**, 1.
Tam, K.K. (1968). *J. Fluid Mech.* **34**, 145.
Thomas, L.H. (1953). *Phys. Rev.* **91**, 780.
Tietjens, O. (1925). *Z. Angew. Math. Mech.* **5**, 200.
Tollmien, W. (1929). Gottingen Nachrichten, Math. Phys. K1., p. 21.
Tollmien, W. (1935). Gottingen Nachrichten, Math. Phys. K1., p. 79.
Van Dyke, M.D. (1975). *Perturbation Methods in Fluid Mechanics*, Parabolic Press, Stanford.
Watson, J. (1960). *J. Fluid Mech.* **9**, 371.
Watson, J.N. (1958). *Theory of Bessel Functions*, Cambridge Univ. Press.

Index

Mathematics and its Applications

Series Editor: G. M. BELL, Professor of Mathematics, King's College (KQC), University of London

Artmann, B. **The Concept of Number***
Balcerzyk, S. & Joszefiak, T. **Commutative Rings***
Balcerzyk, S. & Joszefiak, T. **Noetherian and Krull Rings***
Baldock, G.R. & Bridgeman, T. **Mathematical Theory of Wave Motion**
Ball, M.A. **Mathematics in the Social and Life Sciences: Theories, Models and Methods**
de Barra, G. **Measure Theory and Integration**
Bell, G.M. and Lairs, D.A. **Co-operative Phenomena in Lattice Models Vols. I & II**
Berkshire, F.H. **Mountain and Lee Waves**
Berry, J.S., Burghes, D.N., Huntely, I.D., James, D.J.G. & Moscardini, A.O. **Teaching and Applying Mathematical Modelling**
Burghes, D.N. & Borrie, M. **Modelling with Differential Equations**
Burghes, D.N. & Downs, A.M. **Modern Introduction to Classical Mechanics and Control**
Burghes, D.N. & Graham, A. **Introduction to Control Theory, including Optimal Control**
Burghes, D.N., Huntley, I. & McDonald, J. **Applying Mathematics**
Burghes, D.N. & Wood, A.D. **Mathematical Models in the Social, Management and Life Sciences**
Butkovskiy, A.G. **Green's Functions and Transfer Functions Handbook**
Butkovskiy, A.G. **Structural Theory of Distributed Systems**
Cao, Z-Q., Kim, K.H. & Roush, F.W. **Incline Algebra and Applications**
Chorlton, F. **Textbook of Dynamics, 2nd Edition**
Chorlton, F. **Vector and Tensor Methods**
Crapper, G.D. **Introduction to Water Waves**
Cross, M. & Moscardini, A.O. **Learning the Art of Mathematical Modelling**
Cullen, M.R. **Linear Models in Biology**
Dunning-Davies, J. **Mathematical Methods for Mathematicians, Physical Scientists and Engineers**
Eason, G., Coles, C.W. & Gettinby, G. **Mathematics and Statistics for the Bio-sciences**
Exton, H. **Handbook of Hypergeometric Integrals**
Exton, H. **Multiple Hypergeometric Functions and Applications**
Exton, H. ***q*-Hypergeometric Functions and Applications**
Faux, I.D. & Pratt, M.J. **Computational Geometry for Design and Manufacture**
Firby, P.A. & Gardiner, C.F. **Surface Topology**
Gardiner, C.F. **Modern Algebra**
Gardiner, C.F. **Algebraic Structures: with Applications**
Gasson, P.C. **Geometry of Spatial Forms**
Goodbody, A.M. **Cartesian Tensors**
Goult, R.J. **Applied Linear Algebra**
Graham, A. **Kronecker Products and Matrix Calculus: with Applications**
Graham, A. **Matrix Theory and Applications for Engineers and Mathematicians**
Griffel, D.H. **Applied Functional Analysis**
Griffel, D.H. **Linear Algebra***
Hanyga, A. **Mathematical Theory of Non-linear Elasticity**
Harris, D.J. **Mathematics for Business, Management and Economics**
Hoksins, R.F. **Generalised Functions**
Hoskins, R.F. **Standard and Non-standard Analysis***
Hunter, S.C. **Mechanics of Continuous Media, 2nd (Revised) Edition**
Huntley, I. & Johnson, R.M. **Linear and Nonlinear Differential Equations**
Jaswon, M.A. & Rose, M.A. **Crystal Symmetry: The Theory of Colour Crystallography**
Johnson, R.M. **Theory and Applications of Linear Differential and Difference Equations**
Kim, K.H. & Roush, F.W. **Applied Abstract Algebra**
Kosinski, W. **Field Singularities and Wave Analysis in Continuum Mechanics**
Lindfield, G. & Penny, J.E.T. **Microcomputers in Numerical Analysis**
Lord, E.A. & Wilson, C.B. **The Mathematical Description of Shape and Form**
Marichev, O.I. **Integral Transforms of Higher Transcendental Functions**
Massey, B.S. **Measures in Science and Engineering**
Meek, B.L. & Fairthorne, S. **Using Computers**
Mikolas, M. **Real Function and Orchogonal Series**
Moore, R. **Computational Functional Analysis**
Müller-Pfeiffer, E. **Spectral Theory of Ordinary Differential Operators**
Murphy, J.A. & McShane, B. **Compution in Numerical Analysis***
Nonweiller, T.R.F. **Computational Mathematics: An Introduction to Numerical Approximation**

Ogden, R.W.	**Non-linear Elastic Deformations**
Oldknow, A. & Smith, D.	**Learning Mathematics with Micros**
O'Neill, M.E. & Chorlton, F.	**Ideal and Incompressible Fluid Dynamics**
O'Neill, M.E. & Chorlton, F.	**Viscous and Compressible Fluid Dynamics***
Rankin, R.A.	**Modular Forms**
Ratschek, H. & Rokne, J.	**Computer Methods for the Range of Functions**
Scorer, R.S.	**Environmental Aerodynamics**
Smith, D.K.	**Network Optimisation Practice: A Computational Guide**
Srivastava, H.M. & Karlsson, P.W.	**Multiple Gaussian Hypergeometric Series**
Srivastava, H.M. & Manocha, H.L.	**A Treatise on Generating Functions**
Shivamoggi, B.K.	**Stability of Parallel Gas Flows***
Stirling, D.S.G.	**Mathematical Analysis***
Sweet, M.V.	**Algebra, Geometry and Trigonometry in Science, Engineering and Mathematics**
Temperley, H.N.V. & Trevena, D.H.	**Liquids and Their Properties**
Temperley, H.N.V.	**Graph Theory and Applications**
Thom, R.	**Mathematical Models of Morphogenesis**
Toth, G.	**Harmonic and Minimal Maps**
Townend, M. S.	**Mathematics in Sport**
Twizell, E.H.	**Computational Methods for Partial Differential Equations**
Wheeler, R.F.	**Rethinking Mathematical Concepts**
Willmore, T.J.	**Total Curvature in Riemannian Geometry**
Willmore, T.J. & Hitchin, N.	**Global Riemannian Geometry**
Wojtynski, W.	**Lie Groups and Lie Algebras***

Statistics and Operational Research

Editor: B. W. CONOLLY, Professor of Operational Research, Queen Mary College, University of London

Beaumont, G.P.	**Introductory Applied Probability**
Beaumont, G.P.	**Probability and Random Variables***
Conolly, B.W.	**Techniques in Operational Research: Vol. 1, Queueing Systems***
Conolly, B.W.	**Techniques in Operational Research: Vol. 2, Models, Search, Randomization**
Conolly, B.W.	**Lecture Notes in Queueing Systems**
French, S.	**Sequencing and Scheduling: Mathematics of the Job Shop**
French, S.	**Decision Theory: An Introduction to the Mathematics of Rationality**
Griffiths, P. & Hill, I.D.	**Applied Statistics Algorithms**
Hartley, R.	**Linear and Non-linear Programming**
Jolliffe, F.R.	**Survey Design and Analysis**
Jones, A.J.	**Game Theory**
Kemp, K.W.	**Dice, Data and Decisions: Introductory Statistics**
Oliveira-Pinto, F.	**Simulation Concepts in Mathematical Modelling***
Oliveira-Pinto, F. & Conolly, B.W.	**Applicable Mathematics of Non-physical Phenomena**
Schendel, U.	**Introduction to Numerical Methods for Parallel Computers**
Stoodley, K.D.C.	**Applied and Computational Statistics: A First Course**
Stoodley, K.D.C., Lewis, T. & Stainton, C.L.S.	**Applied Statistical Techniques**
Thomas, L.C.	**Games, Theory and Applications**
Whitehead, J.R.	**The Design and Analysis of Sequential Clinical Trials**

**In preparation*